FineScale Modeler BOOKS

MODELING WORLD WAR II IN THE PACIFIC

Edited by
Aaron Skinner

Kalmbach Media

Kalmbach Media
21027 Crossroads Circle
Waukesha, Wisconsin 53186
www.KalmbachHobbyStore.com

All photos were taken by the respective authors, except IJN cruiser *Tone*, pages 52 and 64; B-25, page 76; Ki-67 "Peggy" bomber, pages 84 and 93; USS *Ticonderoga*, pages 94 and 102; and P-61 Black Widow, pages 120 and 125, © 2020 Kalmbach Media.

Please follow appropriate health and safety measures when working with materials and equipment. Some general guidelines are presented in this book, but always read and follow manufacturers' instructions.

Published in 2020
24 23 22 21 20   1 2 3 4 5

Manufactured in China

ISBN: 978-1-62700-724-5
EISBN: 978-1-62700-725-2

On the Introduction page, from top right, Destroyers alongside USS *Lexington* assist in the carrier's abandonment during the Battle of the Coral Sea. U.S. Navy/National Archives; USS *Oklahoma*, U.S. Naval History and Heritage Command photograph/Captain F.H. Whitaker Photo Collection; Fourth Division Marines come ashore from an LVT amphibious tractor while under fire on Namur Island in the Kwajalein Atoll on Feb. 1, 1944. U.S. Navy

Cover Photos: IJN cruiser *Tone*, P-61 Black Widow, William Zuback; Australian Matilda tank, Andy Cooper

Editors: Aaron Skinner, Eric White

Book Design: Lisa Bergman

Photographer: William Zuback

Library of Congress Control Number: 2019946455

Contents

Introduction

While war had raged across Europe for two years, enveloping the Soviet Union, North Africa, and the Atlantic Ocean, the Pacific region had remained relatively peaceful. That ended on December 7, 1941, when the Japanese navy attacked Pearl Harbor, bringing America into the conflict that was now a true world war. During nearly four years of hostilities, Allied troops fought the Japanese on land and sea across battlefields as varied as the jungles of New Guinea, the tropical beaches of Tarawa, and the frozen islands of the Aleutians. Remote places like Guadalcanal, Iwo Jima, Midway, and Leyte Gulf became household names. While much of the equipment used carried over between the European and Pacific theaters, the latter is characterized by naval combat, especially the elevation of the aircraft carrier to the preeminent warship. Amphibious landings were common during the island-hopping campaigns. Tanks designed to fight in open terrain were heavily modified for the close combat of the jungle. Gathering some of the best modelers from the pages of FineScale Modeler, this book will guide you through 17 projects to build some of the machines from the Pacific War. Happy modeling!

1 Righting a wrong

The recovery of the USS Oklahoma *BB-37*

LUKE EASTER

The USS *Oklahoma* was one of the first victims during the attack on the U.S. Pacific Fleet in Pearl Harbor on Dec. 7, 1941. The first torpedoes struck her hull at 7:56 a.m. She was hit by as many as eight torpedoes in less than 12 minutes, rolled to port, and capsized with the loss of 429 sailors and Marines. Thirty-two sailors were subsequently rescued from the capsized hull. She was a proud and powerful ship that was robbed of her glory that fateful day. For many years, I have wanted to build this 1/350 scale diorama honoring the USS *Oklahoma* and her crew, but my skills as a model builder were not up to the task.

Several years ago, my wife and kids gave me a subscription to *FineScale Modeler*. It was at that point my modeling skills started to grow. At first, I was intimidated by the articles I read. Doing a water scene, well, that was never going to happen! Then, in the October 2006 issue of *FSM*, I read "Making waves" by Tim Reynaga, on his water diorama of a German Type VII U-boat. It was absolutely beautiful and mesmerizing, and everything I needed to motivate me to test the waters. I followed every step in the article, creating my first water diorama. I've since done many water scenes, using everything from Celluclay to resin.

Finally, my skills have caught up with my dreams of building this diorama of the USS *Oklahoma*. I hope I have honored the brave sailors and Marines who served on the mighty dreadnought, and the engineers who so diligently recovered her glory.

This is the moment Luke wanted to capture in his diorama. The righting of a capsized vessel using rotational force is called parbuckling.

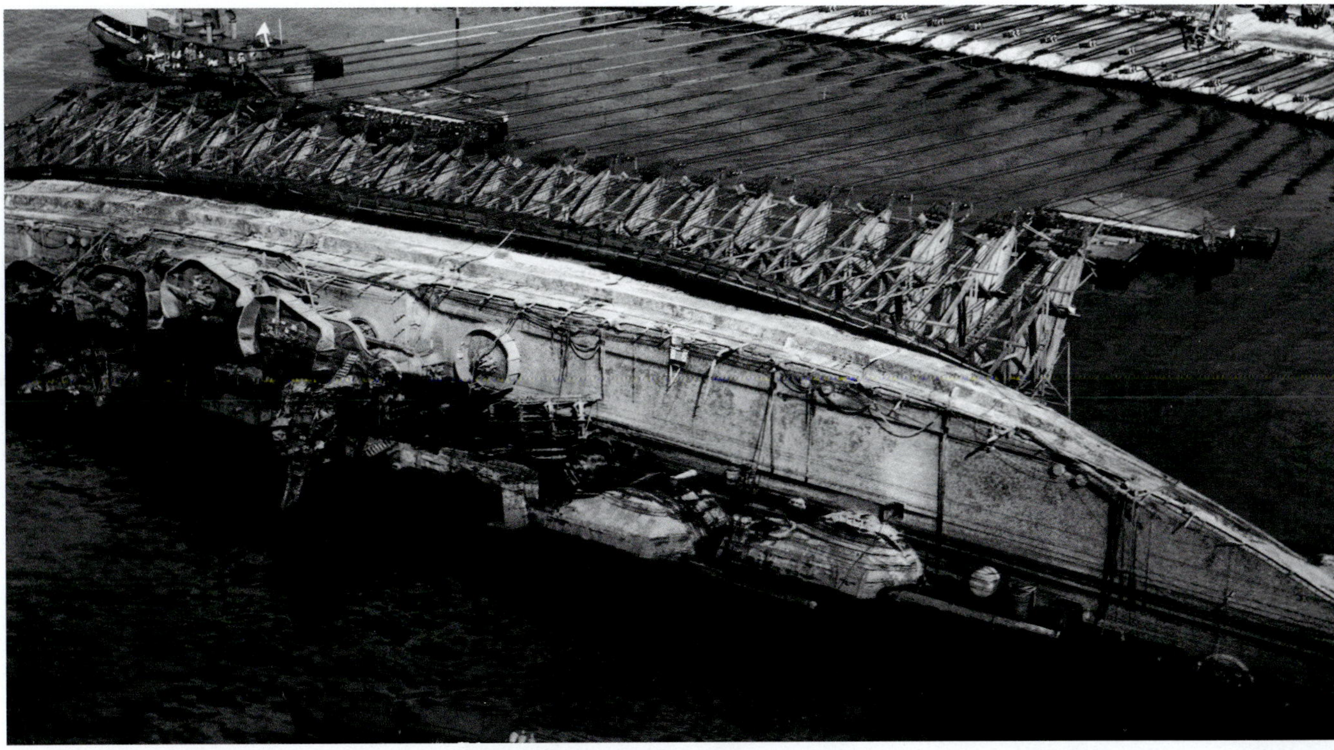

The capsized battleship is seen during the righting process, at the 90° position.
U.S. Naval History and Heritage Command photograph
Captain F.H. Whitaker Photo Collection

I looked at hundreds of historic photos before starting the build. This close-in view of the ship during the righting process shows the Oklahoma at the 68° from horizontal position.

U.S. Naval History and Heritage Command photograph/ Captain F.H. Whitaker Photo Collection

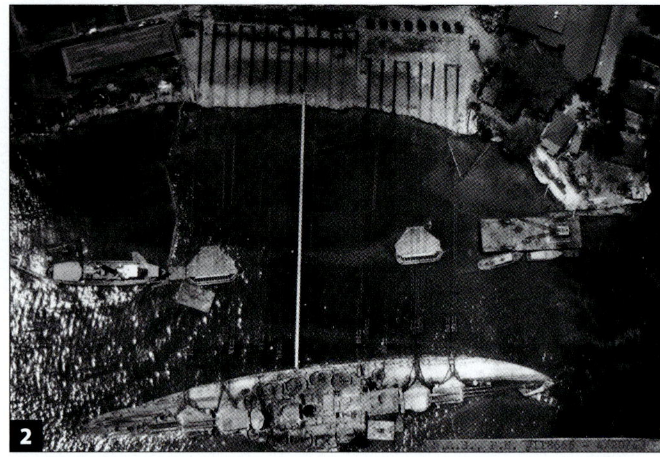

An aerial view of what I wanted to capture, showing salvage operations. The ship is at the 10° position, showing the righting winches constructed ashore.

U.S. Naval History and Heritage Command photograph/ Captain F.H. Whitaker Photo Collection

The project began with construction of the supporting details. These included the 17 derricks that transferred the force of winches on shore to the hull of the ship. A simple flat jig made of pins and a sheet of foam helped ensure all of the derricks were the same.

I made an angled jig to finish the derrick assembly.

This derrick is nearly finished. It just needs the pulley on top.

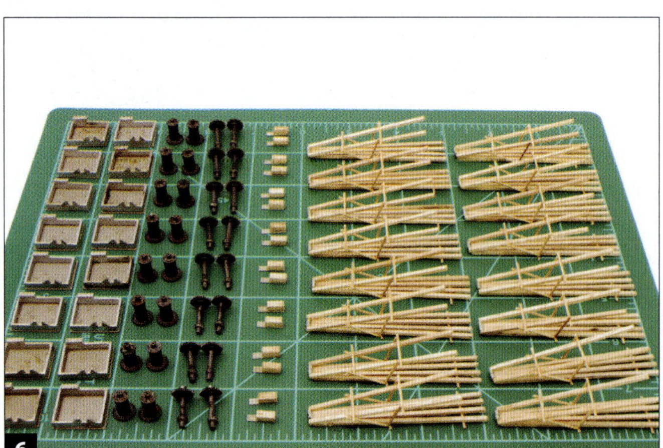

I worked in subassemblies to build all of the infrastructure for my *Oklahoma* diorama. From left are the winch boxes, winch mechanisms, electric control boxes, and derricks.

To build the winches, I used watch gears, buttons, and wooden dowels. Again, I made the parts assembly-line style.

8

Concrete and steel boxes housed the winches and anchored them to the shore. This declassified photo shows righting winches 1 through 12, constructed on Ford Island. These winches were used to retract cables connected to the capsized hull of the battleship, in order to right the ship.

U.S. Naval History and Heritage Command photograph/Captain F.H. Whitaker Photo Collection

9

I built the winch boxes from styrene strip. The electric box is styrene and balsa. I used Berkshire Junction EZ Line for the cables.

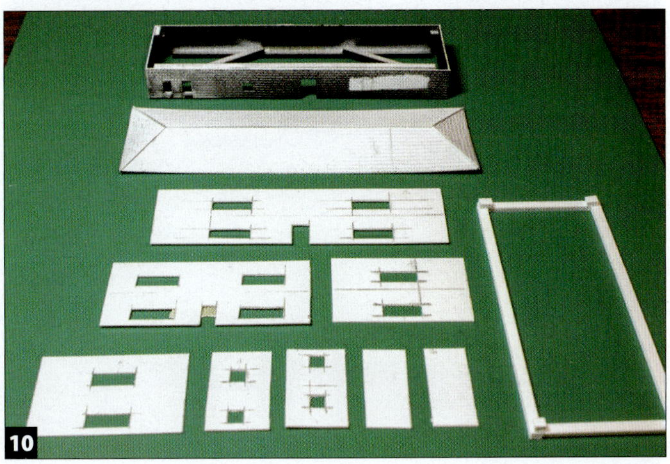

10

I scratchbuilt the dorm and office building based on historic photos using Evergreen styrene.

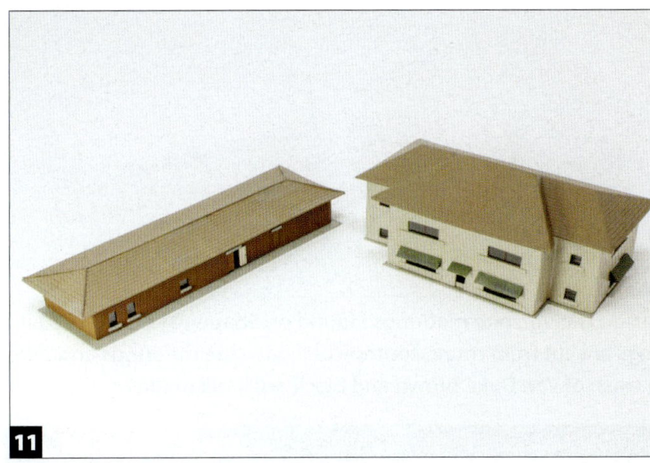

11

Here they are completed and ready for the diorama.

12

I used a Black Dog House No. 1 accessories set for the project.

13

This U.S. Navy barge was scratchbuilt from Evergreen styrene strip and sheet, then given an artist-oil wash of raw umber, Van Dyke brown, and black. The Caterpillar crane is from Nikko Models.

14

The launches were detailed with photo-etched (PE) railings from Gold Medal Models and the HobbyBoss USS *Arizona*.

15

The storage tanks were made from the leftover bottles of desiccant that came in my medicine bottles. I used leftover PE railings and ladders to add detail.

16

I used USS *Arizona* moorings I found on Shapeways.com. The pilings are cut from round toothpicks. I gave the moorings and piles a wash of Van Dyke brown and black artist oils.

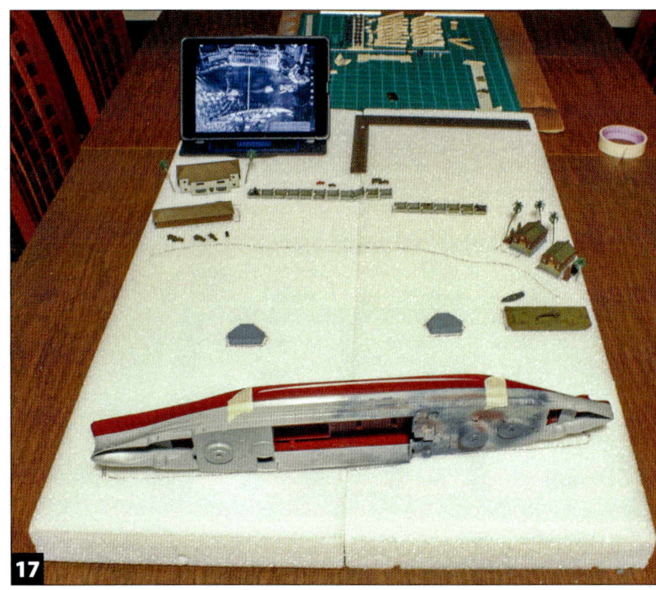

17

I used a historic photo on my iPad to help me lay everything out to scale.

18

To make absolutely sure, I had the photo enlarged at a print shop to ensure I got everything truly to scale. I taped the hull of a HobbyBoss USS *Arizona* together for size, and placed several of the completed structures.

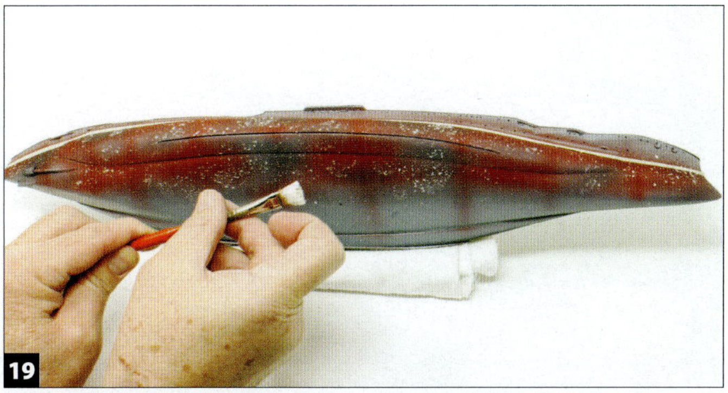

19

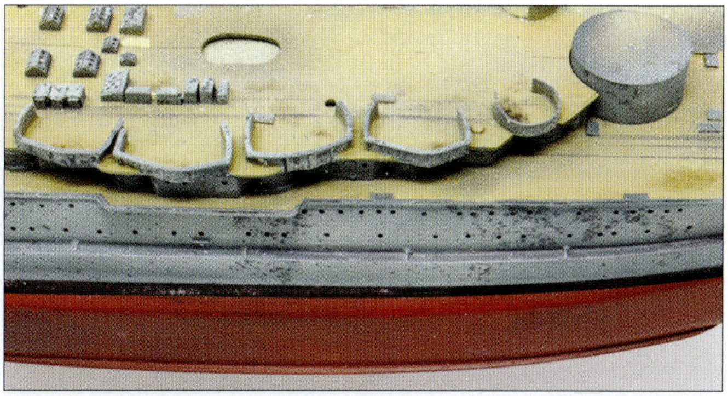

For my model of the *Oklahoma*, I used Iron Shipwright's 1/350 scale kit. I pre-shaded the bottom of the hull with Model Master flat black and Humbrol matt red (153). Once that dried, I used a medium-sized brush and blotted the hull with water. I then poured salt randomly over the hull. The salt and water coagulated and dried.

Then I mixed plaster of paris with water to form a runny mixture, which I flicked from a brush with my finger, spattering plaster randomly over the hull. I post-shaded the hull with Humbrol matt dark red (73). When that was dry, I wiped the salt and plaster away with a damp sponge, creating a heavily weathered hull.

20

The same technique was used above the waterline, except I pre-shaded with Tamiya German gray (XF-63) and post-shaded with dark sea gray (XF-54). The effects are easy to see on this turret.

21

I completed the weathering with a heavy artist-oil wash using raw umber. This created a heavily weathered and battered USS *Oklahoma*.

22

This ship's crane is seen in almost every historic photo of the *Oklahoma*.

U.S. Naval History and Heritage Command photograph/Captain F.H. Whitaker Photo Collection

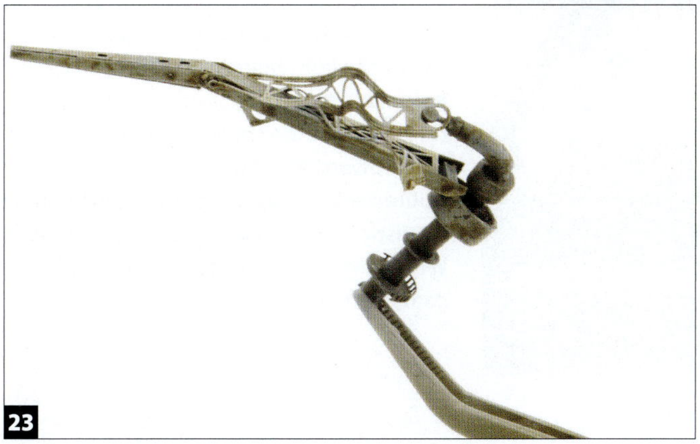

23

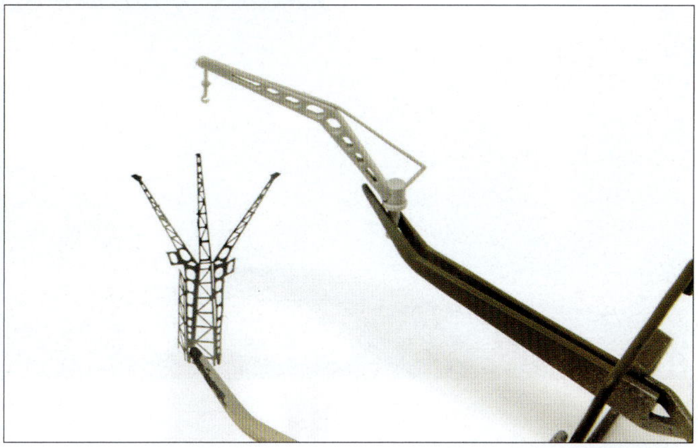

I opted for the Gold Medal Models USS *Arizona* PE aircraft crane, left, over the one provided in the Iron Shipwrights USS *Oklahoma* kit.

The vehicles came from L'Arsenal.

I made the trees using products from Woodland Scenics and JTT Trees.

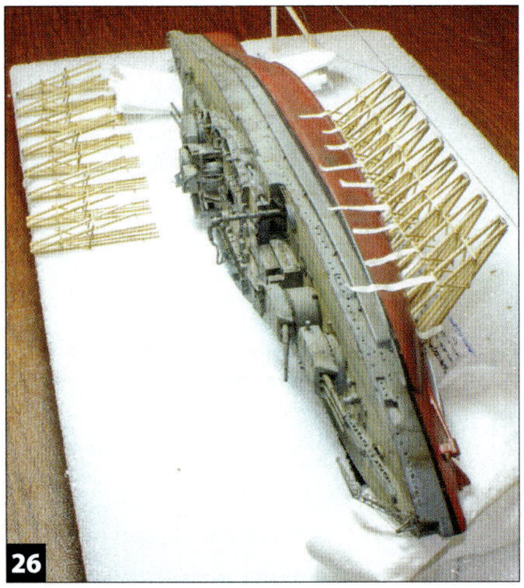

The engineers made the derricks very straight, so I used some EZ Line and a jig to match their efforts.

Looking forward on March 8, 1943, the Oklahoma is in the 130 degree position. Her starboard deck edge is just rising from the water.

U.S. Naval History and Heritage Command photograph

This shows the waves coming ashore and the concrete and steel boxes that held the winches. The ocean floor was airbrushed in phases. I started at the shoreline using a light beige. I then cut random small patterns in masking tape and put it on top of the light beige, then sprayed dark beige to create sand spots. I then used turquoise for the shallow water and faded back toward the ship with darker blue. I poured several layers of Woodland Scenics Realistic Water, then made the waves with Woodland Scenics Water Ripples and a coarse brush.

An early war SB2U Vindicator

2

Special Hobby's 1/72 scale kit helped fit this Midway bomber into a crowded showcase

FRANK CUDEN

Frank couldn't help but notice the similarity of the SB2U cowling/forward fuselage area to the Kingfisher, with the aft third of the fuselage and tall tail similar to that of the BT-13.

Before the advent of the Accurate Miniatures kits, I came upon the HiPM 1/48th scale kit of the SB2U Vindicator at a model contest vendor table and immediately bought it for the grand sum of $20. With every intention of building it, I winced a bit when Accurate Miniatures released its superb Vindicator kits. However, I still was intent on having a go with the rather crude HiPM kit. But with showcase space dwindling because of new additions, and with Special Hobby releasing its 1/72 scale Vindicator kit, No. SH72264, I just knew I had to build the smaller one instead to save space, and so I relegated the HiPM kit to the back of the shelf, and will most likely donate it to a club raffle one of these days.

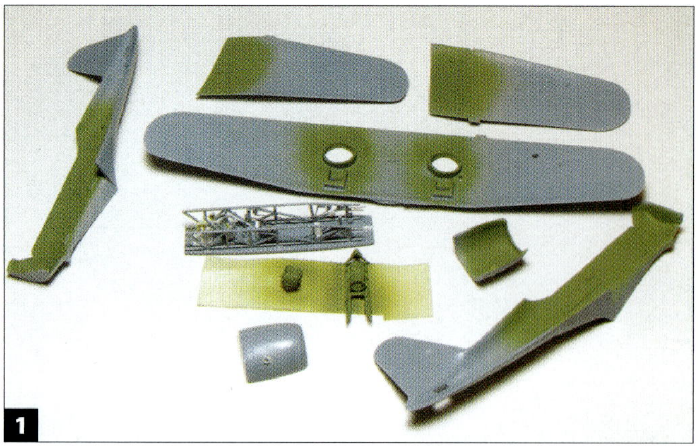

I began to assemble the interior, airbrushing as I built. The cockpit walls are of the "cage" construction type. The Squadron In-Action book on the Vindicator inspired scratchbuilt details added to the various kit parts in the cockpit assembly.

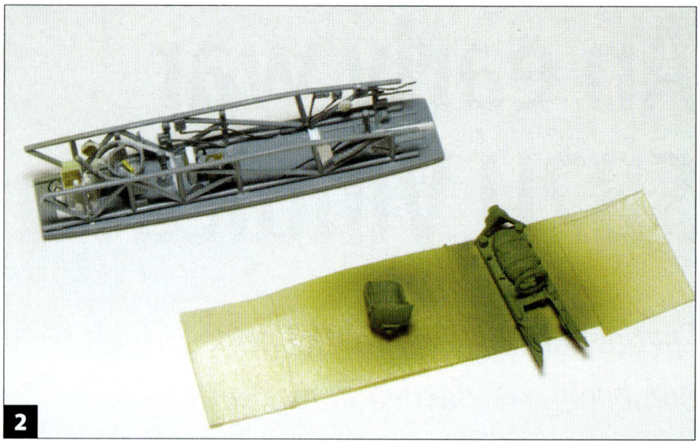

One can see additional cockpit items, particularly the Apoxie Sculpt life raft I made to replace the tank-like molding provided in the kit. Seat belts and other placards came from ReHeat photo-etched metal(PE) sets I had in my parts box.

With cockpit wiring, the directional antenna, and paint work, the cockpit is pretty much complete. The homemade raft is strapped in place on the turtle deck.

Everything fit quite well into the fuselage half. I left the life raft assembly off of the model until later, not wanting it to get in the way.

I glued the fuselage halves together along with the wing, then added putty to hide seams.

An intake part fits at about the 10 o'clock position on the cowl. I painted black to add depth behind the intake. I also added reinforcement blocks for the cowl and engine. To break up the interior color, I painted the pilot's seat aluminum. Seat belts and shoulder harnesses came from the ReHeat set.

7

Most of the major assembly is complete. Looking at the model at that point, it seemed to sit too high in front, so I carefully cut sections from the landing-gear legs, then I glued them back together with a piece of wire inserted in the center for reinforcement. I now had a more accurate Vindicator stance.

8

A good portion of the SB2U was covered in painted fabric, the rest was metal. I used two shades of Testors Model Master enamel, light gray (No. 36495), the lighter of the two shades for the metal areas; and light gray (No. 36492), a slightly darker shade for the fabric areas. I established panel lines by lightly going over them with an artist's soft-lead drawing pencil.

9

With minimal decals to be applied, I was quickly through with that chore. The difference between the lighter and darker shades is evident. Note the addition of the wheels, belly tank, and life raft turtle decking. I lightly dry-brushed the tank with Tamiya Acrylic flat aluminum which gave it more of a natural metal look after it had been sprayed with Alclad II aluminum.

10

A close-up of the cockpit shows the scratchbuilt life raft, which I would discover is obscured under the canopy, detail wiring added to the turtle decking just in front of the circular gunner's seat, and a small circular head rest for the pilot made from a section of plastic rod. Note also the black walkway decals as well as the hard-to-see white "7's" just outside of them.

11

I added the machine gun next, set in an offset raised position. I later found PE parts for the sight ring and "pipper" so they were added later. White metal gun belts were painted brass and copper. I marked every fourth bullet red, to indicate a tracer round, and installed it on the left side of the machine gun.

12

To improve the kit canopy, I cut it into sections and vacuum-formed each segment. I installed the windshield first with thinned white glue, then added each section to complete the opened canopy arrangement. The rear-most canopy section slid forward and then tucked on an angle under the aft canopy section, just forward of the gunner's compartment.

13

I continued adding details, such as the PE sighting ring and "pipper" on the machine gun. As luck would have it, I installed, and then cracked off and lost, the antenna mast just forward and to the left of the cockpit. I made a new one and thankfully did not lose that one. Thin fishing line was used for the antenna wire with tiny drops of white glue as insulators.

14

The kit provided the three-color prop tip decals and Hamilton Standard logos. I masked and sprayed the backs of the tips yellow. Having forgotten to drill a hole in the windshield to accept the telescopic gunsight tube, I held my breath and successfully drilled a small hole through the glued-in canopy. I drilled out each end of the gunsight to add depth, then glued it in place.

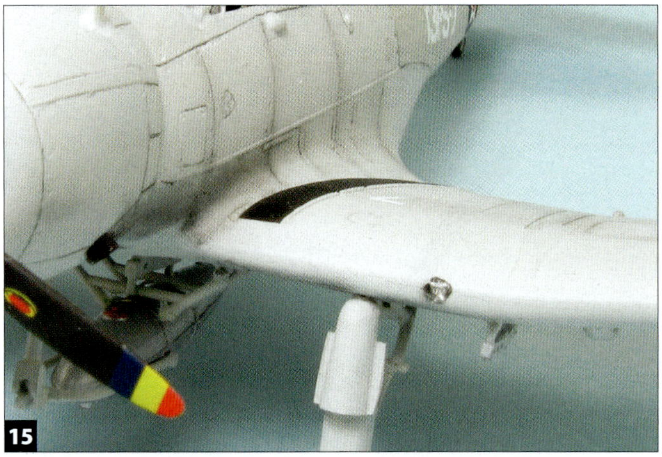

15

I tried a different technique to replicate the landing lights. I cut a small square of aluminum foil slightly larger than the hole, and gently stuffed it into the hole. I secured the foil in place with white glue, and when that was dry, I placed a drop of Tamiya clear over the foil, bringing it level with the surrounding wing area.

16

I drilled away the molded-on exhaust pipes and added new ones made from plastic tubing. I used my reamer set to thin the plastic on the inside, giving a scale appearance to the stacks. In addition, the bomb/belly tank cradle gave me some fits as I worked with the parts, but in the end, I succeeded, getting everything plumb and secure.

17

I highlighted some panel lines with gray pastel along with black, brown and dark gray pastel smudges to simulate exhaust staining. Using a small brush and a little pastel dust, I lightly dragged the brush along the panel lines and any build-up was easily blown off. If the build-up was too heavy, I wiped the excess off with a cotton swab. I added a black wash to the prop hub to accentuate the detail.

18

For me it's next to impossible to achieve a straight edge when brush-painting, so I used short strips of black decal to stripe the tail hook, providing crisp edges rather than the jagged ones hand-brushing would have produced.

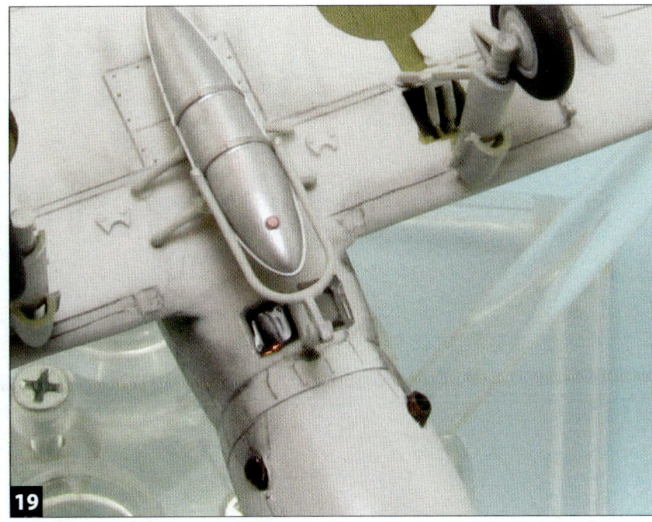

19

Two ventral observation window holes were cut out just aft of the lower cowling and I filled them with Microscale Kristal Klear.

20

While not the easiest of builds, including vague areas on the instruction sheet, the kit went together well and turned into a nice replica of the early World War II aircraft. The plane lends itself well to a couple of other war time color scheme options in the kit so those wanting a 1/72nd scale SB2U Vindicator in their collections can easily be steered in Special Hobby's direction. Come to think of it, the Special Hobby Vindicator is the ONLY one these days, in 1/72nd scale. Try it—you'll like it.

3 A Matilda for the jungle

Australian forces found modified British Matilda tanks were effective against the Japanese in the Pacific

ANDY COOPER

The Matilda tank had made a name for itself in the campaigns of the North African deserts. However, by 1941 it was being outclassed by the increasing pace of desert warfare in terms of speed and firepower.

The Matilda seemed destined for training and other non-frontline roles, but by mid-1942 the situation in the Pacific was becoming dire for the Allies with Japanese forces threatening mainland Australia. The need for tanks in New Guinea to stem the Japanese advance saw around 200 Matildas initially sent to Australian forces, who subsequently found the vehicles well-suited to jungle warfare.

During operations with the Australian forces, their Matildas were continuously modified to suit conditions and operational requirements as the campaigns unfolded, giving them some unique characteristics over their British counterparts.

Australian forces operated several versions of the Matilda: the standard armed 2-pounder, the 3-inch howitzer-armed close support (CS), and a flame thrower known as a Frog where the flame gun was mounted in the turret in place of the main armament. There was also a version with

The Tamiya kit is a good base to start with, but to get that unique Australian look for the model any of the Firestorm resin sets for the Aussie Matildas is a must. The set was well cast and provided all the main modifications mentioned plus a host of other bits and pieces to add stowage.

mortars mounted on the rear known as a Hedgehog.

My model is a Matilda CS of the Australian 2/4th Armoured Regiment operating on Bougainville in May 1945.

To build my Aussie Matilda, I used the Tamiya No. 355 "Red Army" Matilda. The 1/35 scale Tamiya kit is a good base to make an Australian version from as it includes the "spudded" tracks fitted to Australian tanks, and though not mentioned in the instructions, also includes the 3-inch howitzer gun barrel if you decide to make the Australian close support (CS) version.

For the unique Australian bits and pieces, I used Australian manufacturer Firestorm Models kit No. FS35205, which provided a turret protection ring, engine deck mesh or PSP (pierced steel planks) protection and armored track guards, plus numerous stowage items typically seen on jungle-warfare Matildas.
Star decal sheets 35-C1068 and 35-C1069 gave me a choice of markings to choose from for 2-pounder gun tanks, 3-inch howitzer close support, and even a dozer-equipped version.

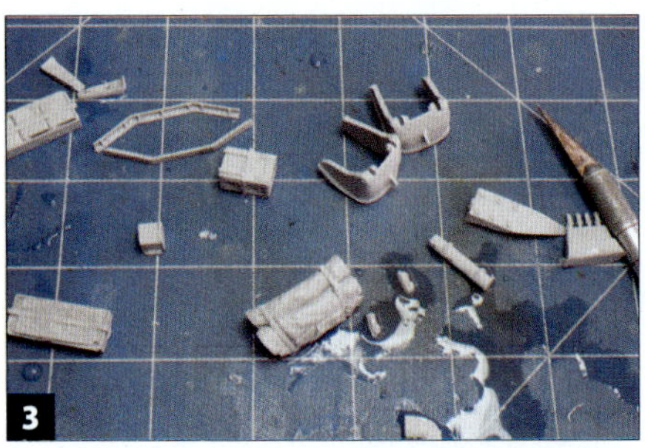

First job was to clean up the Firestorm resin. I dipped each part in water before removing it from its pouring stub with a small saw. Making the parts wet keeps dust to a minimum, but wearing a mask is advisable as resin dust is harmful. Once the parts were removed from the stubs, they were clearly marked with the part number where necessary. With one more wash in mild soapy water to remove any remaining residue, the parts were left to air dry and then bagged up.

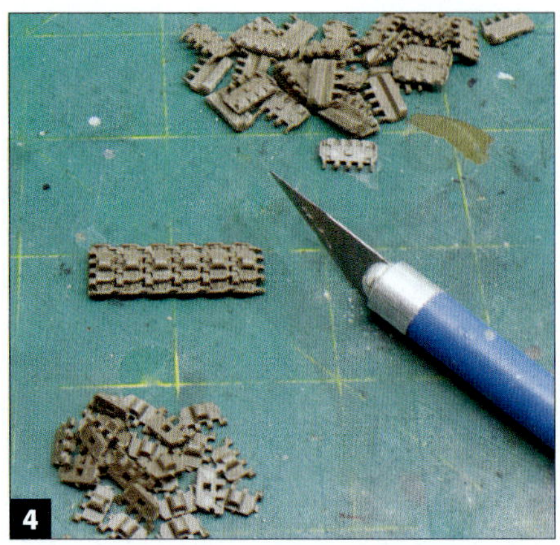

The Tamiya Matilda comes with individual track links. Each link consists of two parts, and the tank requires 138 individual track links (69 per side). I broke down the task by making sections of 10 links at a time while listening to music—not a hard task, just a little mind-numbing at times.

The guide horns on the Tamiya tracks are molded as a solid piece whereas they are hollow on the real thing. There are aftermarket tracks available (Bronco AB3531 for example) but considering very little is seen of the tracks, especially the guide horns, on the finished model I stuck with the kit-supplied tracks.

Tamiya approached the glacis plate assembly in a clever way to eliminate the usual seam in such areas. The plate comes in three parts, the joints being crafted along the weld seams.

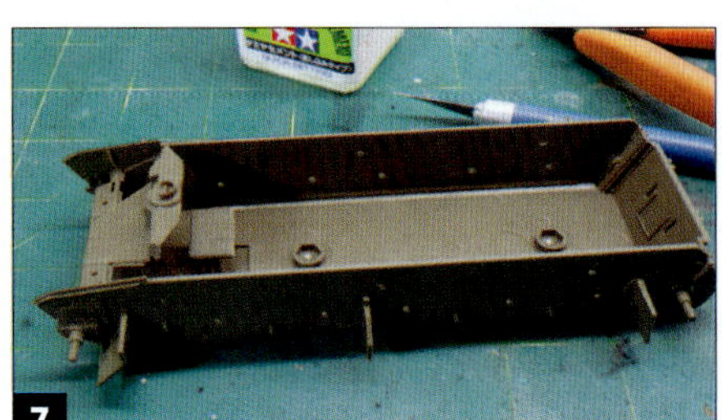

With the glacis plate glued together, the remainder of the lower hull was quickly finished and work began on the running gear.

The running gear is simple and easy to assemble, and as it is all but hidden behind the armored skirts any lack of details perceived or otherwise is not an issue.

9 The drive sprocket and front idler pop into place on the provided poly caps, which makes it easy to adjust positioning when fitting the tracks.

10 The mud chutes and upper track guides need to go on before the tracks are fitted, and again, there are no problems here.

11 Not being a fan of "working" tracks, I fixed mine in place by fitting them and brushing on liquid cement to bind them to the running gear.

12 Despite having the same number of links each side, the right hand side needed a link removed to make it fit. I think this was more of my doing somehow than a Tamiya fault. I positioned the gap above the sprocket so it would be hidden behind the armored skirt, fixed the rest of the track in place, then cut a spare link down to fill the gap.

13 I primed and painted the running gear and side skirts before completing the hull, though in hindsight this was not really necessary. The Australian Matilda's were "jungle green." I used Tamiya NATO green. The shade is probably a little light, but will be toned darker with washes and weathering.

14 The Australian modifications from the Firestorm resin kit started with fitting the turret protection ring while ensuring the turret still turned within it. Once placement had been marked with a pencil, I secured the turret ring parts in place with superglue.

15 The Firestorm track guards portray perfectly the prominent heavy cast feature of these additions. I fixed the four small pivot mounts into the guards (parts No. 8) before mounting them onto the tank. The two guard locks (parts No. 9) can then be glued into place once the guards are attached.

16 Many period photos showed no headlights fitted, so I filled the slots in the hull top with scrap styrene, then shaved it flush with a hobby knife.

17 I restored the cast texture with Tamiya putty mixed with a little liquid cement. I brushed this over the area, then stippled it with an old, stiff brush.

18 Photos made it clear no two tanks were the same. My goal was a "typical" Australian Matilda of the period. I scratchbuilt and added track guard locks, and a bar across the front of the hull to hold stowage. Firestorm resin items were positioned accordingly.

19 While the Red Army Matilda instructions have you use the raised cupola, the majority of photos show Australian Matildas equipped with the lower cupola. Fortunately the lower cupola is included in the Tamiya kit (part D50).

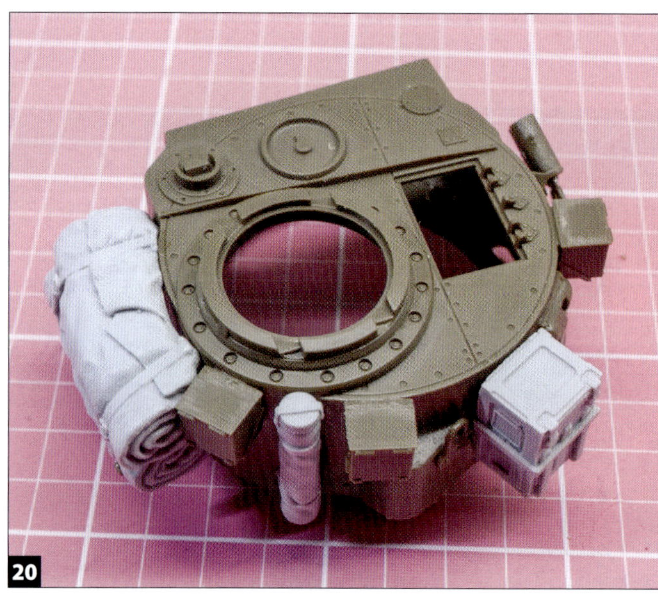

20 The turret can now be completed with the resin additions and the main weapon of choice, in my case the 3-inch howitzer barrel, which is not optioned in the instructions for the "Red Army" version, but is part D12

21

The antenna for the "Red Army" version is incorrect for Australian tanks. The correct parts are D2 and D7. You will need to drill holes in the turret top, one in the small square plate forward of the loaders hatch (for part No. 2), and the other behind the loaders hatch (for part 7). The locations are molded on the inside of the turret. I would advise leaving the smoke grenade thrower off until you are finished handling the turret.

22

After a few sessions, the Matilda had received most of its resin and scratchbuilt modifications. The Firestorm engine deck mesh was trial-fitted at this point along with the one-piece resin rear load, but I removed them before painting the tank so they could be painted separately.

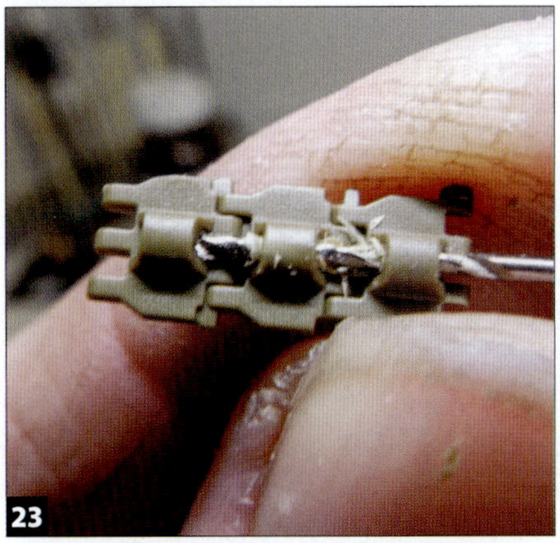

23

As mentioned earlier, the Tamiya track guide horns are molded solid. As the spare track links on the glacis plate are clearly visible, I drilled out the guide horns to create a hollow appearance.

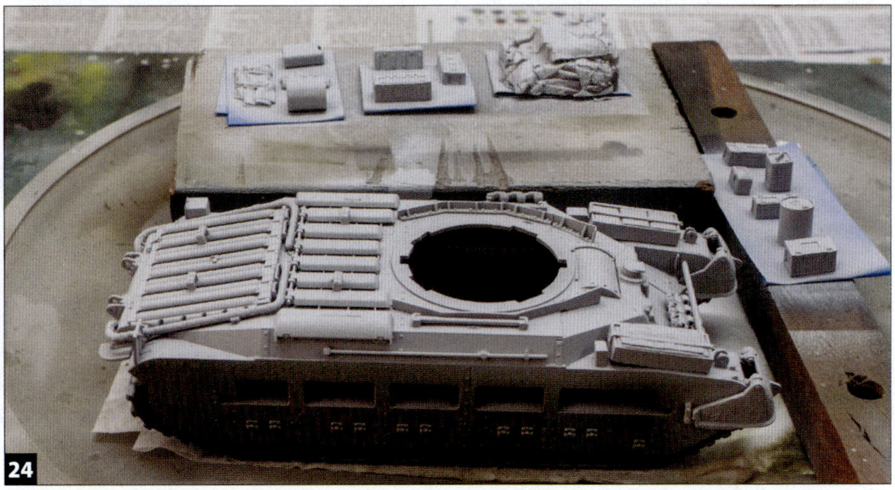

24

The model and resin load parts went to the spray bench for a coat of Tamiya gray primer. I sprayed the paint from a bottle mixed with lacquer thinner, then left the parts to dry.

25

The Australian Matildas were marked on the flanks with red fording depth stripes. I sprayed both sides with insignia red in the appropriate places. The lines were then masked with 2.5mm-wide strips of Tamiya masking tape. I went across the gaps of the mud chutes and then trimmed the excess tape off with a fresh scalpel blade.

26

Some Australian Matilda's had a white air recognition symbol painted on the turret. I thought this would break up the overall green scheme, so to achieve this I sprayed the turret top flat white.

27

Using a compass cutter, I cut a 38mm diameter circle from masking tape, then I cut another inside it measuring 19mm diameter.

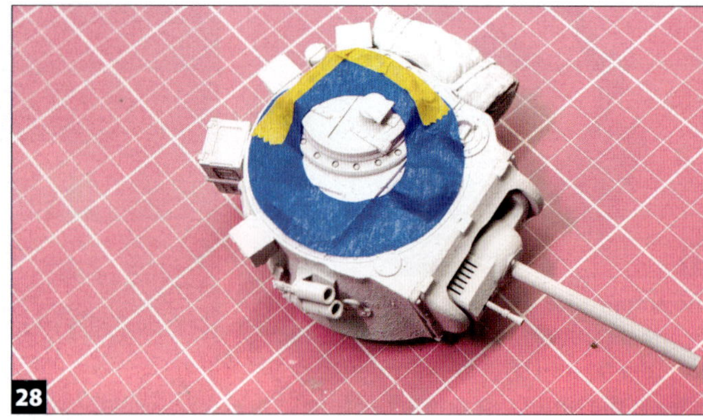

28

I applied this to the turret top over the many lumps and bumps. The mask needed some adjustment with Tamiya tape. My research suggests these markings were field-applied, and as such would not be supremely accurate. A little latitude here is probably acceptable.

29

Finally it was time to apply some green. I used Tamiya NATO green, spraying the entire model. After the first coat had dried, I went back with the airbrush and added some different shades of green to vary the finish, with darker shades lower on the hull and lighter shades nearer the top. Once all this had cured I removed the masking and hand-painted various details with acrylics. When I was happy with the effect,

30

I sealed the model with Alclad II Aqua Gloss coat in preparation for the decals. Decal solution helped settle the turret numbers over the cast texture. Once the decals were in place the model was resealed with Aqua Gloss Clear using an airbrush.

31

For the next stage, I applied an artist-oil wash of raw umber, let it sit for half an hour, then carefully wiped away excess. When this process was complete I used oil pinwashes of burnt sienna to imitate rust and general grime and some white to replicate water run off. These washes are applied with a small brush as a dot, then streaked downward with a thinner-soaked brush to create subtle streaks of varying opacity.

32

Earlier test-fitting of the resin one-piece load for the engine deck had shown that it needed to be hollowed out on the underside to sit correctly. I did this with a motor tool (wear an appropriate mask), rinsing between grinding and test-fitting until it sat correctly.

Finally, the mesh was superglued into place and the load placed on top and glued down. I used thread to make tie-down ropes. I added more stores from the Firestorm conversion kit, then painted everything with acrylic colors. Once the load had dried, it was sealed and weathered as described for the tank itself.

Jungle Matildas frequently became caked in mud and torn foliage, much of it falling off the tracks and down the mud chutes. To create this look, I made a mix of white glue and appropriate water color paint. To add bulk, I added fine sawdust. The foliage within the mud is dried herbs from the kitchen cupboard.

I selectively applied this mix with an old brush, using period pictures as a guide. Before the mix dried, a pinch of the dried herbs was sprinkled onto the wet mix and the excess shaken off.

Once all this was dry, these areas were given color depth by mixing dark earth and chocolate brown paint together, then carefully and randomly spraying it with the airbrush. When this was dry, the entire model was given its first flat coat.

The final touch was to apply gloss sealer to some of the muddy patches to simulate areas that were still wet. This brought my Aussie Matilda build to completion.

4 Resin upgrades for a Wildcat

The 1/72 Hasegawa kit benefits from Quickboost and True Details sets

VLADIMIR KAFKA

There is no need for a long introduction of the F4F Wildcat. It took part in all of the most important battles in Pacific campaign, from Coral Sea and Midway through the Guadalcanal campaign to the heroic fight of the "Taffy 3" group in the Leyte Gulf.

Early in the war, it had a hard time against the much more nimble Zeros, but after pilots adopted improved tactics, this sturdy fighter more than stood its ground, ending up with a highly positive kill ratio.

While this camouflage is simple, it is more than compensated for by the well known, thrilling story of Captain Henry T. "Hammerin' Hank" Elrod, who was posthumously awarded the Medal of Honor for his feats during the defense of Wake Island in the first week of the war. Elrod was the first pilot to sink a warship, the destroyer IJN *Kisaragi*. His light caliber bombs struck the depth charges on the stern of the ship, setting them off.

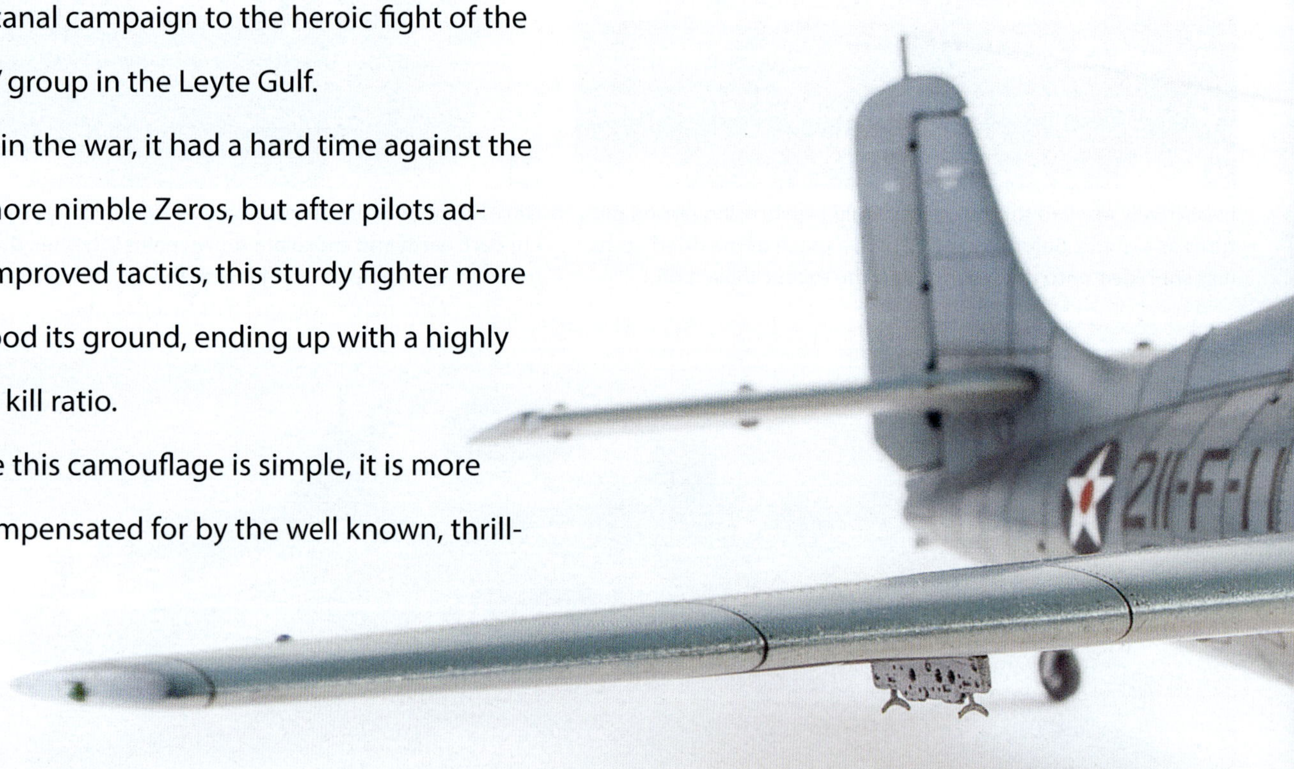

Wake Wildcats were brand new, delivered just a couple of days before the Japanese attack and actually were freshly repainted during the deployment, to comply with the latest camouflage directive. Engines and propellers were masked with tarps, which overlapped the cowlings a bit, leading to a non-standard camouflage line typical for this squadron.

I chose the old Hasegawa model over the Airfix kit for its more delicate panel lines and surface details, and a more accurate shape of fuselage top behind the cockpit.

One thing I don´t like about the Hasegawa kit is its canopy, which is visibly small, and flush with the fuselage, which would make it impossible to slide backward. I solved this issue with a small aftermarket item of my own making.

Given the bland camo and no weathering, I had to focus solely on a clean build and nice details to have a remote chance to impress anyone with this model. While the Hasegawa kit itself is excellent and would be trivial to assemble on its own, properly incorporating all the aftermarket parts—mainly the resin wheel wells and cockpit—took some effort. I think the result has been worth it and I like it, simple as it is.

Wildcats at Wake Island were brand-new, and on top of that were freshly painted to comply with recent camouflage directives. Vladimir shows that a simple paint scheme can still build an interesting model, especially with the addition of a few detail sets.

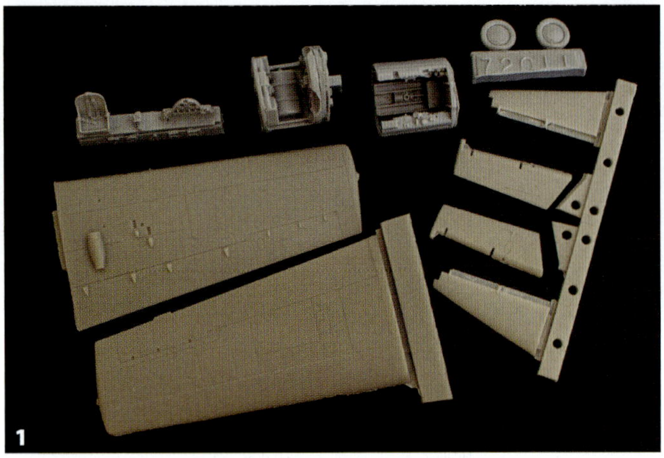

1 I gathered all the aftermarket pieces in one place. The amount of resin raises a philosophical question: Is this a plastic kit with resin accessories, or a resin kit with a plastic fuselage?

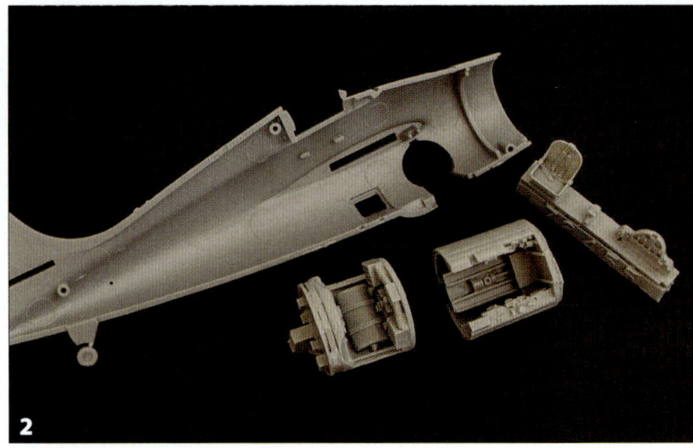

2 The Hasegawa kit celebrated its 25th anniversary in 2019. Its biggest downsides are its cockpit, which is very plain, and wheel well, which is nonexistent. Both issues can be addressed at once with True Details' aftermarket set: cockpit and wheel well are cast as two almost complete assemblies, ready to be joined together and inserted into the fuselage.

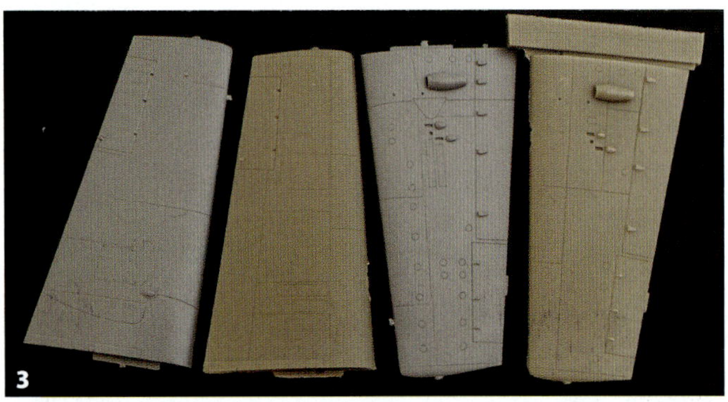

3 The other downside is that Hasegawa only really ever made an F4F-4 with -4 version wings: folding, with 6 guns. Quickboost comes to the rescue with resin wings for the -3 version: non-folding, with 4 guns. A comparison of kit parts (light gray) and the resin wings (tan) shows the panel lines are obviously different. There is one goof worth mentioning, too: note the gun holes in leading edges don't align with their respective gun panels, they seem to be shifted exactly one interval.

4 Elrod's Wildcat wouldn't be complete without its light bomb racks. I took them from an F3F-2 kit by Special Hobby (it's probably safe to assume they were identical on the F4F). It would be cool to include the bombs as well, but I was concerned I wouldn't manage to maintain their proper attachment and geometry via those flimsy photo-etched metal (PE) cradles.

5 The engine is that proverbial "generic Hasegawa radial," but in a kit already overloaded with resin, it does its job. Anyway it's half hidden in a relatively narrow cowl, behind intercooler intakes and bulky propeller cuffs. Here, it's already sprayed with Gunze aluminium (C8).

6 I added copper ignition wires, glued into appropriately sized holes ...

7 ... then bent over the cylinders and secured with dabs of superglue. (no need for clean work as it's hidden under the cowl). Everything here is painted with Vallejo paints.

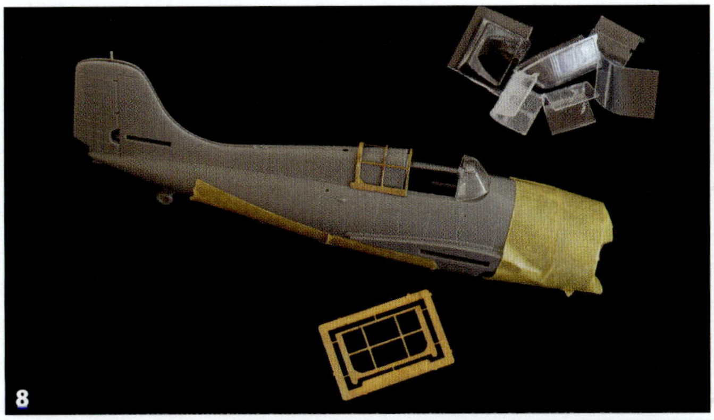

8

The canopy out of the box seems to be slightly undersized. It is exactly flush with the fuselage top behind the cockpit, which means in reality you couldn't slide it backward. The windshield looks a bit anemic, too. I bought a Squadron vacuum-form set, intended for Academy kit. In contrast, these parts are huge. At least the windshield is usable after some trimming both in shape and size. For the sliding part, I drew a canopy frame that I later had photo-etched. I now sell these under my Airone Hobby brand. In the upper right corner you can see a "graveyard of ideas."

9

There is a wide recess in front of the cockpit to accommodate the original windshield part. I filled it with thick superglue and sanded the glue. Note the cockpit partially painted interior green. Later, I realized the proper shade should be bronze green ...

10

... a much darker, colder shade, without the yellowish tint. The tubular parts of the True Details set are already joined together. I removed the blanked rectangles from the bottom of the cockpit casting behind the bombing windows in the bottom of the fuselage, extending the cutouts because the thick walls of the resin part would be visible from outside through the windows.

11

Here is the top side of the cockpit. I really didn't like the set being cast in an almost completely assembled form (minus instrument panel and seat). This made painting more difficult. The wheel well was the same in this regard.

12

Later, I changed my mind and replaced the True Details seat as I didn't like belts cast in place, with a not very well defined result. The replacement is a plastic part from Sword's P-40 (a bucket seat is a bucket seat after all) with colored PE belts from Eduard's F6F Hellcat.

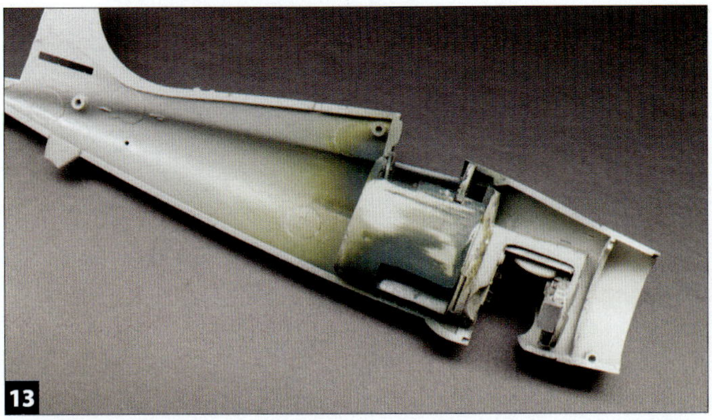

13

The fit of the resin assembly in the fuselage was OK, but not perfect. I found either I got the geometry wrong in the cockpit part, or the wheel wells, but never quite both.

14

The gaps were mainly in the cutouts in the rear bulkhead (these gave clearance for the wheels). Repeated filling and sanding took place. Note the drilled out exhausts. I decided not to replace the kit parts, even though they look slightly inaccurate.

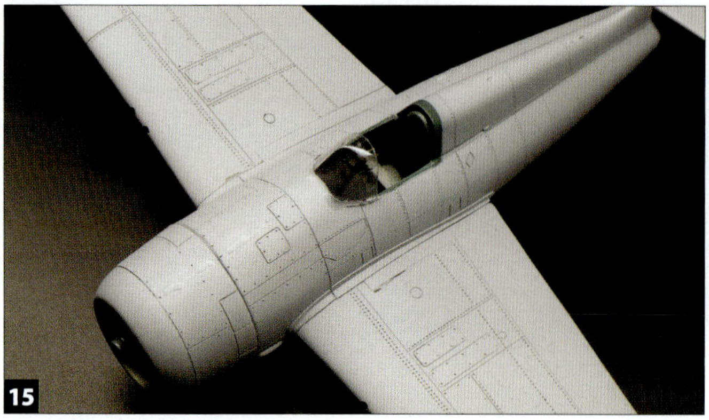

15 I attached the windshield with superglue after testing to be sure it wouldn't fog. I used thick surfacer primer to smooth the connection to the fuselage, along with other imperfections such as the fuselage seam. Using a sewing needle, I added holes representing the 100 or so fasteners on the front fuselage and engine cowl.

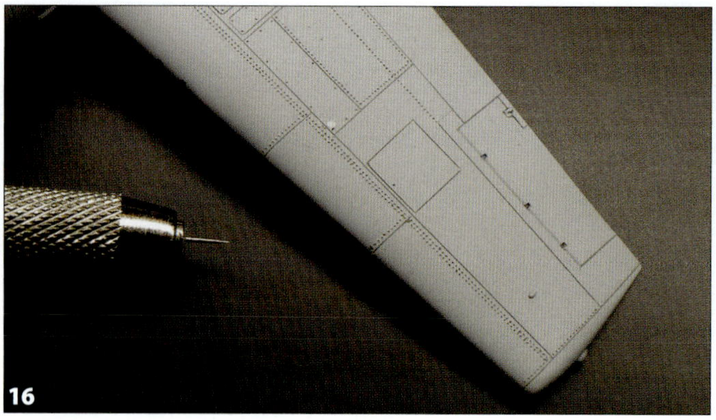

16 Riveting into resin is a bit more difficult than into. Photos of the full-size plane show a big difference between rivets on the wings and fuselage. While the fuselage has prominent lines of raised ballhead rivets, the only riveting visible on the wings is a double line going along the main spar. I added a few other lines with my needle here and there as well though, as a compromise between realism and aesthetics. Note that I sanded down molded ribs on the control surfaces; I'll re-create them with shading later.

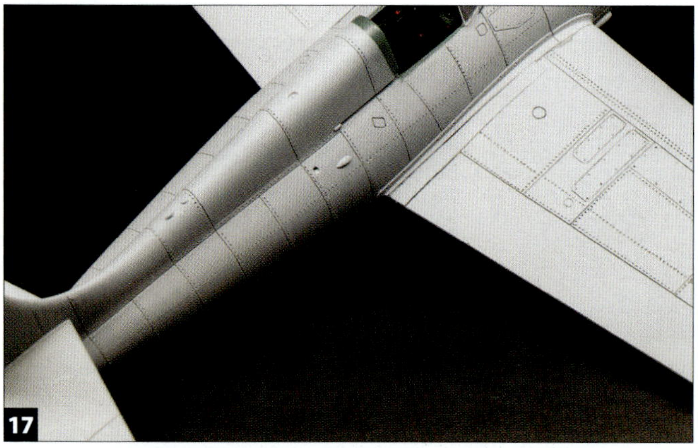

17 I riveted the fuselage after assembling the halves and attaching both wings. In hindsight, this was a mistake, as maintaining straight lines along the barrel-shaped fuselage was challenging. Again, I used thick surfacer to correct goofs.

18 Having no masks for the windshield, I masked each frame one by one. This is a far lesser evil than trying to mask everything at once with carefully measured tiny strips. Having two airbrushes really helped here: one with base black color, one with gray-blue. I used leftover paint to check for imperfections here and there, giving the model a patchwork appearance.

19 The result is not bad, despite visible steps at the frame intersections where the masking overlapped. Note the extra frame in the middle of the side panels of the windshield, a feature distinguishing a -3 version from a -4 version.

20 Much fine sanding and polishing took place before I was satisfied with the surface quality. There were some rough spots on the resin wings, and I added more everywhere while fixing my sloppy riveting.

21 An interesting feature, unique for Wildcats from Wake Island, is the high demarcation line on the cowling. I sprayed this free-hand.

22 This is the PE canopy frame I sell under the brand Airone Hobby. The transparent material is polypropylene film (fake cellophane). The film I used had the added benefit of not being fogged by superglue. Gluing it in is a bit fiddly, but on the other hand, there is no masking to be done, so it took less time in the end. Such crisp frames are impossible to achieve with any other method.

23 I test-fitted the canopy. Speaking of the clear parts, I must say it is still apparent the windshield belongs to a different model. I probably should have done a better job of blending it.

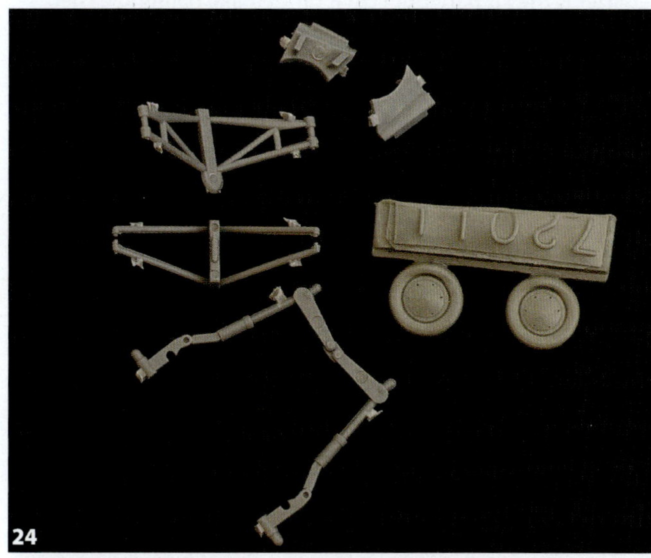

24 Resin replaced the kit wheels, which were marred by ejector-pin marks. The strange oval shapes at the end of the main landing gear legs replicate chains (like on a bicycle) which retracted the whole contraption. They were removed in the next step.

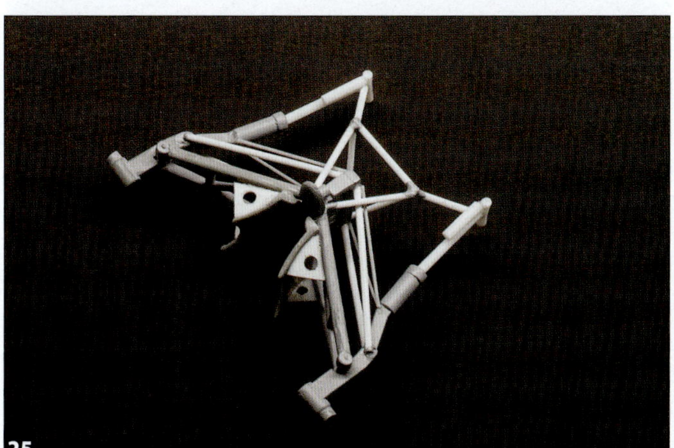

25 I added struts and retracting arms made from stretched sprue and Evergreen styrene shapes. I also replaced the upper sections of the main legs as that was easier than removing the mold seams from the kit parts. The kit's wheel bay doors were as thick as manhole covers, so I thinned them down a lot, and added their front and rear sides, complete with drilled out lightening holes.

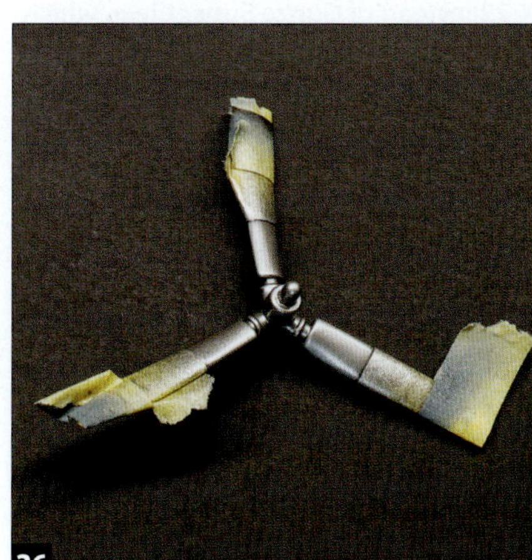

26 The propeller with its bare metal cuffs was sprayed with Gunze super fine silver (SM01). I made the blade tips yellow instead of tricolor. Even though it is contrary to the contemporary directive, Photos show a vast majority of F4F-3s had yellow tips.

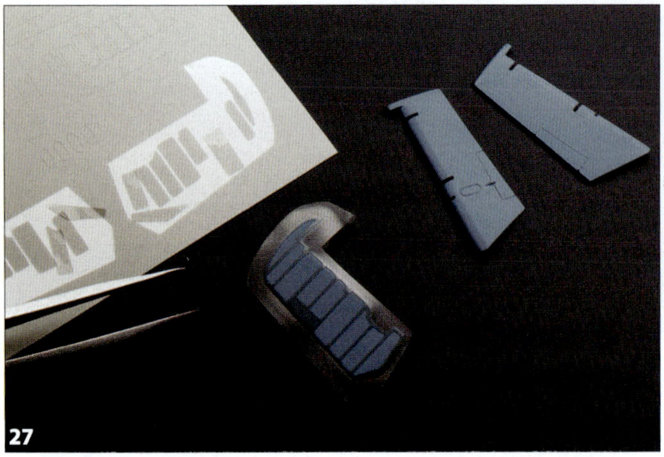

27

I shaded the control surfaces using Airone Hobby masks. They are made of Oramask vinyl film, which is easy to work with, and it is low tack, so adjustments can be made without the risk of ruining the paint. The elevators are resin replacements; the vertical stabilizer is the kit part, but I cut it off to make painting easier and also to better model its attachment to the rudder.

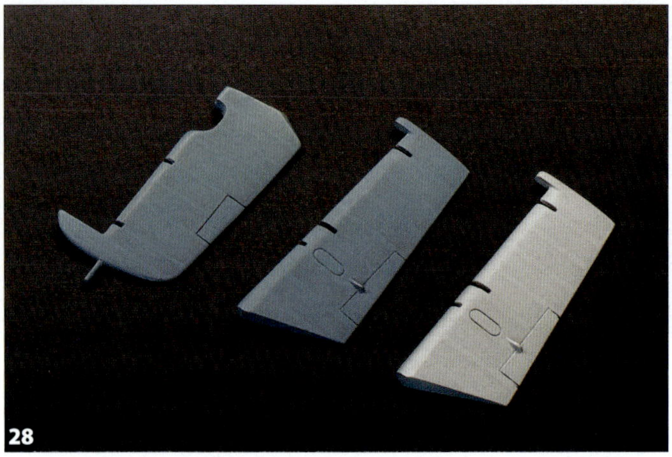

28

I use a mix of black and red-brown, diluted to the consistency of "dirty thinner," then gently sprayed along the mask edges. On a single color camouflage, I found it is actually better to overdo it a bit, focusing just on shape, and then tone the intensity down using a light mist of the base color.

29

With assembly complete, I fought with the clear varnish a bit. I used Tamiya clear gloss X-22 instead of Gunze Super Clear, which I usually use. I polished the surface some more, not because I was bothered by the orange peel surface (to be later hidden under clear flat anyway), but because of some rough spots you can see on the wings. The surfacer (Mr. Surfacer by Gunze) doesn't adhere to resin very well, so when sanding it, instead of creating the usual smooth transition, there was some peeling resulting in rough edges.

30

I followed with washes to pop surface detail, then proceeded to decals. I like the Yellow Wings decals color and perfect register, but they are thick and different decals solutions had wildly divergent effects—from nothing using Tensol, Gunze Mr. Mark Softer Neo or Tamiya Mark Fit Strong, to utter destruction using Hypersol. I sanded away the destroyed decals and replaced them. Fortunately, the decal sheet allows building a whole squadron, so there were a lot of spares. These were freshly painted machines, so apart from some exhaust staining on the bottom side, no weathering was added. Following flat varnish, I called it done.

Mission: Resupply!

5

Building Fujimi's pint-sized cargo sub in a realistic diorama setting

TIM REYNAGA

A lesser known story of World War II is the Japanese submarine effort to sneak emergency supplies to starving imperial army troops trapped on Pacific islands cut off by U.S. forces. These so-called *mogura* ("mole") operations took place in waters dominated by the U.S. Navy and were extremely hazardous, with deliveries taking place only under cover of night. I've always been intrigued by these desperate missions, so when Fujimi released a *Maru-Yu* class cargo transport sub it went right to the top of my to-do list.

These purpose-built cargo subs weren't very big, and the 1/700 scale model is only 2¼ inches long. The tiny submarine would have been lost on a shelf among my other 1/700 scale ships, so placing it in a larger diorama made sense. It could also help tell the tale of these remarkable craft. I decided to depict the little transport unloading its goods somewhere in the Philippines, anchored close to the jungle with troops taking the precious supplies ashore in their Daihatsu landing craft.

Desperation led the Japanese to develop cargo submarines that could run resupply missions to soldiers trapped on islands occupied by Allied forces. Tim built a diorama for the diminutive sub so it wouldn't get lost on his display shelves, and also to explain its role.

REFERENCES

The Japanese Submarine Force and World War II by Carl Boyd and Akihiko Yoshida, United States Naval Institute Press, 1995

"Imperial Japanese Army Transport Submarines: Details of the YU-2 Class Transport Submarine YU-3," by Lieutenant Mark L. Bailey, Royal Australian Navy, *Warship International Vol. XXXV, No. 1* (1998) ISSN 0043-0374

"The 'Maru-yu' Submarines," by Yasufumi Kunimoto, *Warship International Vol. XXXVI, No. 3* (1999) ISSN 0043-0374

Photos of a captured Maru-Yu submarine at Navsource online: www.navsource.org/archives/08/08455.htm

SOURCES

Battlefleet Models
P.O. Box 473318
Aurora, CO 80047-3318
battlefleetmodels.com

Evergreen Scale Models
18620-F 141st Ave. NE
Woodinville, WA 98072
evergreenscalemodels.com

Gold Medal Models
1412 Fisherman Bay Road
Lopez, WA 98261
goldmm.com

Woodland Scenics
P.O. Box 98
Linn Creek, MO 65052
woodlandscenics.com

Fujimi's *Maru-Yu* cargo submarine kit is accurate but basic, so I made a few improvements. (**1**) The main change was to replace the center portion of the sub's deck with a new piece made from .020-inch Evergreen scribed plastic sheet, which better simulates the spaced planking. Also, since the purpose of these subs was transport, I opened up the casing to show off the two unsealed cargo hatches. The hatches themselves were sheet plastic discs cut using a .059-inch Waldron Model Products punch, then hollowed out. Other changes included adding tiny open casing covers, bollards, and cutting away the hull below the waterline. The landing craft, which came from Pit Road's Equipment for Japan Navy Ship-WWII No. 3 (SW-900), received plastic vents, cabins, hose reels, and photo-etched metal (PE) steering wheels. The submarine received a coat of *tsuchikusa iro* ("earth grass color"), a brown-khaki paint commonly used by the Japanese army. Now it might seem odd for submarine to be painted army colors, but these extraordinary vessels actually were built and manned by the Japanese army—completely independently of the imperial navy! The color was my best-guess mixture of Model Master military brown (FS30117) and olive drab (FS34087) enamels. After applying the rising sun decals to the sail, the models were set aside and I turned my attention to the base.

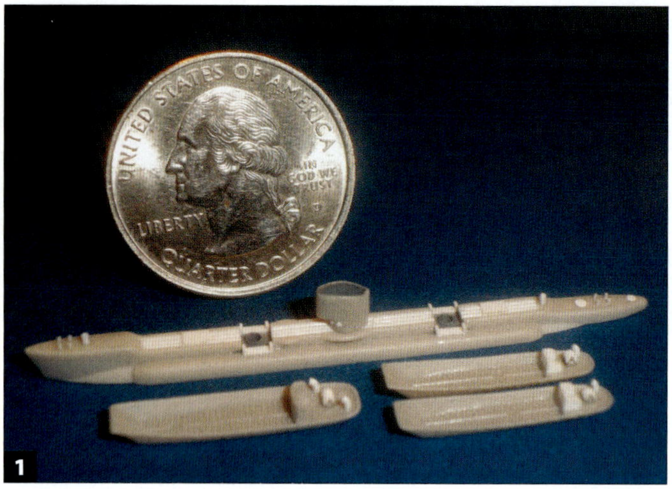

1

2

Completing the initial ocean and groundwork for the scene was simplicity itself; just a layer of Prang Das modeling clay, a nontoxic, air-drying material for children's craft projects. The dense clay is easy to work with, solid but smoothing nicely under water-dampened fingers. The only features in the calm sea were the gentle ripples of the sheltered tropical lagoon. In addition, I lightly pressed the sub into the soft clay to give the model a natural "sit" in the water. The land was even simpler. I wanted to imply deep water by having the ground rise fairly sharply at the shoreline, but there was no need to expend much effort texturing it. Since it was to be covered with forest anyway, I just molded the basic shape.

3

4

5

The tropical jungle was to be a major element of the diorama, but where do you get 1/700 scale trees? Fortunately, small-scale trees aren't that difficult to make. Preserved gyp mini (baby's breath), used for dried flower arrangements, makes great trunks and branches, and Woodland Scenics forest green foam bushes (FC-148) have just the right color and consistency for the leaves. I began with the trunk and branches. The preserved baby's breath is a great material to work with, soft and pliable yet relatively tough. The first step is to cut a piece into a length about equal to the height of your tree, in this case about 2½ inches.

Next, trim away those out-of-scale buds and paint the bare trunks and branches. (**4**) I used Krylon flat brown spray paint, but any suitable brown color would do.

Then comes the fun part, the leaves. (**5**) The Woodland Scenics foam is designed to simulate HO to N (1/87–1/160) scale bushes, so the material as is looks too big for 1/700. To bring it down to scale, simply tear the green foam material into smaller bits and attach the scale "leaves" to the branches with white glue.

6

The resulting trees are petite yet surprisingly durable. (**6**)They can be multiplied to create a fairly simple, cool-looking jungle with brown branches peeking out here and there under the dense green canopy.

7

The sea in this scene is quiet, almost flat, so giving it depth had to be done with color. After smoothing out the Prang Das clay into placid undulations, the water areas received a coat of Model Master non-specular sea blue (FS35042) enamel. I lightened it slightly near the tan-painted shore, but retained the dark color around the sub to suggest deeper water. I also added variations with lightened and darkened mixtures of the basic blue to give the calm waters some interest, but kept the overall contrast subtle and the colors dark to impart a somber feel.

8

After securing the sub and boats to the base, I attached the more delicate details to them. This detailing would normally have been done before fixing the models to the base, but in this case they were so small that the base served as a useful handle so I could work on them without causing damage. The *Maru-Yu* was fitted out with Gold Medal Models "ultra-fine" IJN railing (set 700-25). These special parts are a extremely delicate, but I like them because at .003-inch thickness they are even finer than typical .005 inch PE. The 37mm gun just forward the conning tower (the real ones were taken from army tanks!) was adapted from USN 20mm Oerlikon gun parts. In addition, the sub also received guards, jump wires, cargo handling derricks, periscopes, and other details fabricated from various thicknesses of wire.

9

For a little extra realism I also attached tan-painted wire mooring lines to both the Daihatsu landing craft alongside the sub and to the unoccupied landing craft tied off on the bank. Details such as this, though ordinary, are a simple way to add interest to dioramas.

10

While the supporting elements are important, the visual focus of this little diorama is the life-saving supplies and the soldiers hauling them ashore. (**10**)The provisions are from Battlefleet Models BFM-7103 assorted pallet loads. The imperial army troops are PE brass naval figures from Gold Medal Models (set 700-17). Though many of the tiny soldiers may be difficult to spot, I used over a dozen of them on the boats and along the jungle trail. Re-posed and painted in army green, they drive the action and provide a human sense of scale.

11

The scene is a quiet one, but the soldiers and supplies, along with the sea, jungle, boats and of course the submarine itself, come together to convey the hushed drama of the covert resupply operation.

12

Displaying really small models such as this *Maru-Yu* to their best advantage can be a challenge. The model may be tiny, but with a realistic background and carefully constructed details, even a 2¼ inch sub can help tell a compelling story.

Painting a Pacific pilot

6

A few tricks can go a long way toward making your figures more realistic

JOE HUDSON

This is a 120mm resin figure from Verlinden Productions. Verlinden, which has been out of business a few years now, was ahead of its time when it came to resin kits and the now popular 120mm scale.

Beautifully cast with minimal cleanup and easy assembly, this figure is a good representation of the U.S. Navy pilots who served during the Pacific war.

While researching this figure, I saw that the box art was taken from a picture in the book *U.S. Navy Uniforms In World War II Series Volume 2: US Naval Aviation Flying Clothing and Gear* by Jeff Warner (2007 Schiffer Military Books).

The figure came in eight parts, and the only issue was the cast-on cigar had broken off. It was not in the kit, so once all the parts were removed from the resin carrier blocks, my first step was to create a replacement.

I assembled the kit using superglue, then primed it with Tamiya sky gray (XF-19) applied with an airbrush.

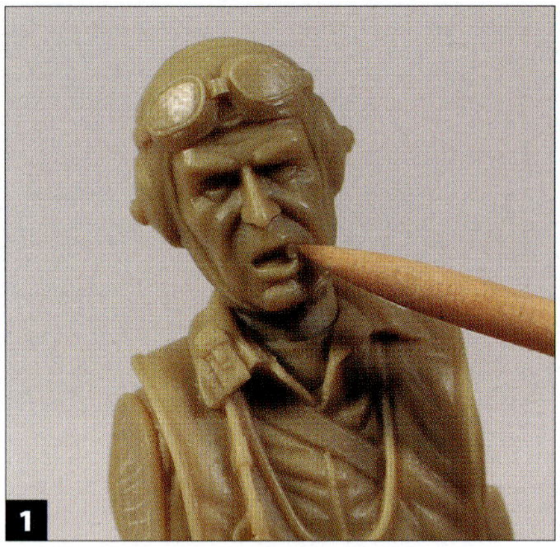

1 A cigar in the pilot's mouth had broken off at some point and was no longer in the kit's box.

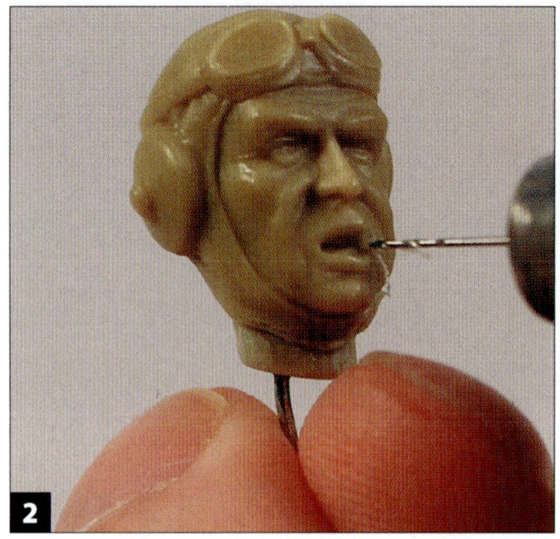

2 Using a small drill bit, I drilled out the broken off cigar.

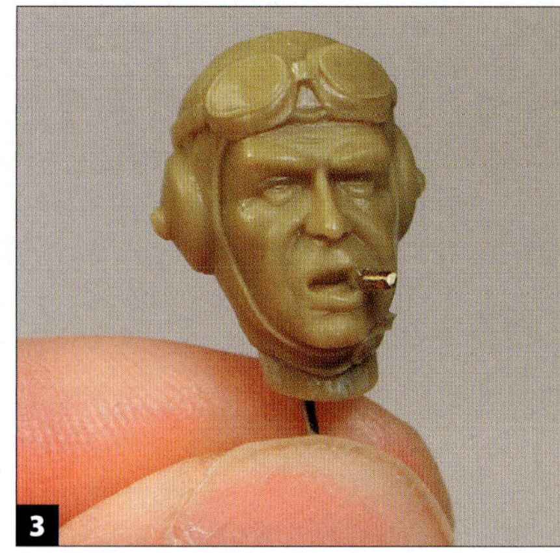

3 I used a small piece of brass rod to create a replacement, supergluing it into place.

4 I painted the face has with a base coat mix of Vallejo Nocturna reddish flesh (74.003), Nocturna base flesh (74.004), and Model Color beige red (70.804).

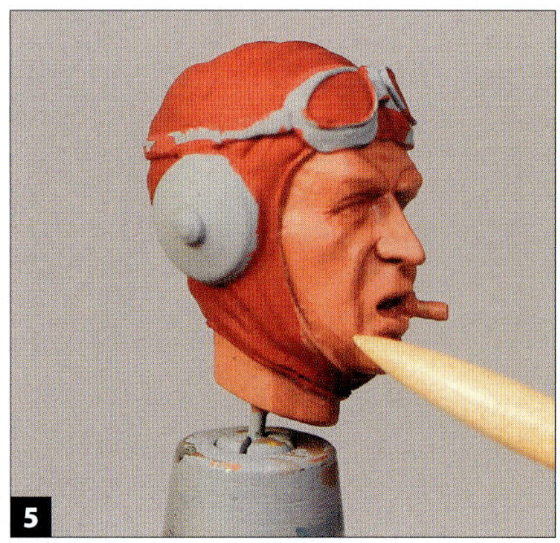

5 I began adding shadows to the face using the base mix with a little Scale 75 Scale Color Indian shadow (SC23).

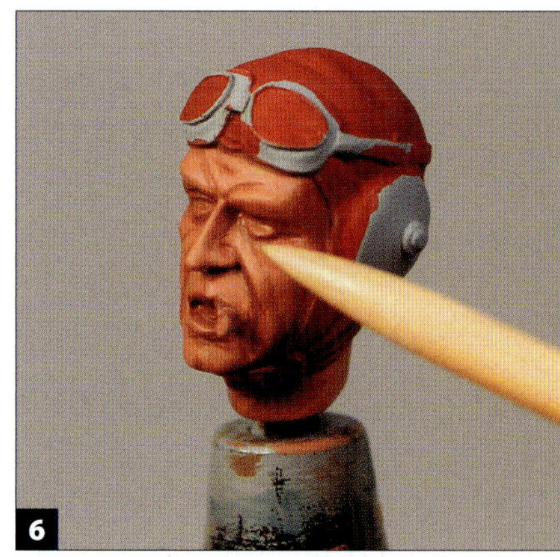

6 I added several layers, each time adding more and more Indian shadow to the base mix for the deepest shadows.

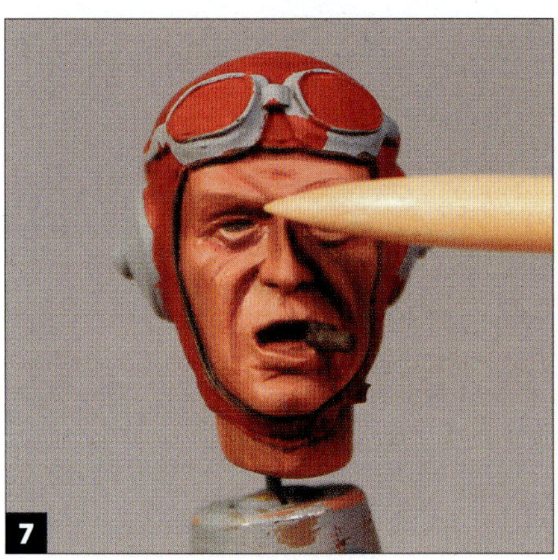

7 To do the eyes, I used Vallejo chocolate brown (70.872) to paint the whole eye socket.

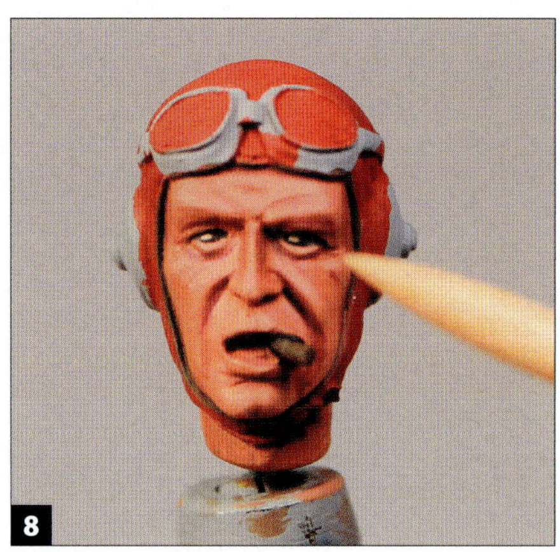

8 I then used Vallejo light skin tone No. 70.928 to paint the corner of each eye to replicate the whites of the eye, leaving a small area of chocolate brown as the pupil. I always try to have my eyes looking either left or right to avoid the 1,000-yard stare. I then took a green color and painted it over the brown to simulate green eyes.

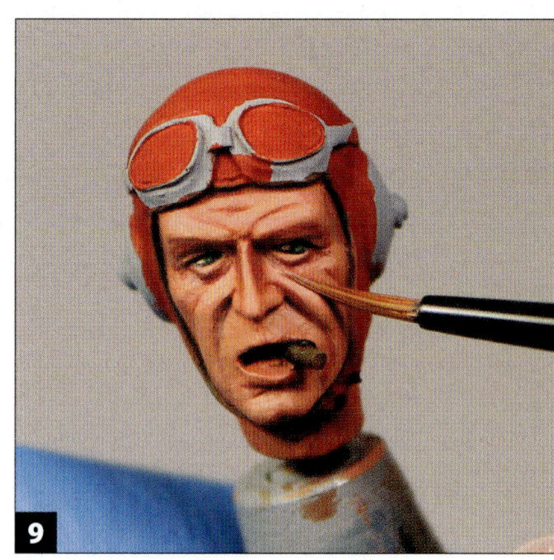

9 I added highlights to the face using the base mix plus more beige red.

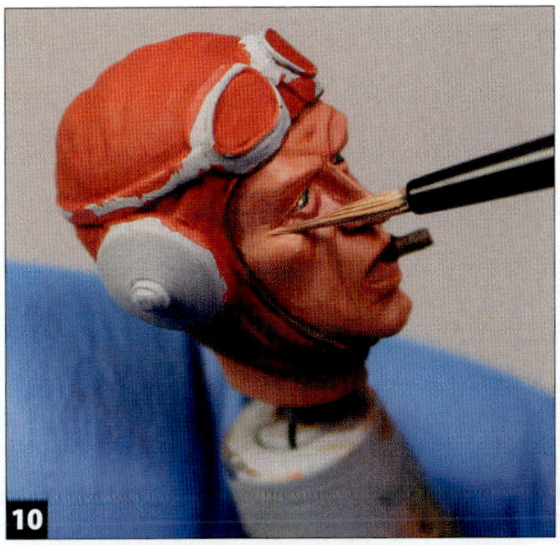

10 I continued to add more beige red until the color was just beige red.

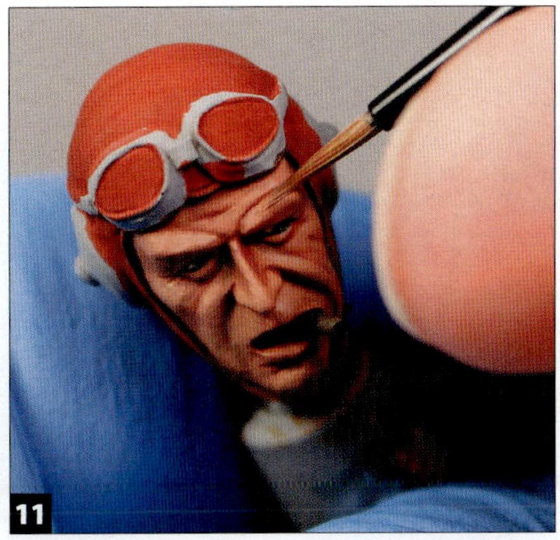

11 I then added some Vallejo sunny skin tone (70.845) to the beige red for the highest highlights.

12 I also took some chocolate brown and went all the way around the face along the helmet to frame the face and give me a better look at the work I was doing.

13 I use photos to check my progress. Once I finished the face, I looked to see if there were any glaring errors or items that I need to work on before moving on to another section of the figure.

14 I painted the red helmet using Scale 75 reds. The colors are listed in 15. I started with a base that was pretty dark, and by doing this I did not have to go back and add medium shadows because of the base color. I did add a deep shadow, but that was all the shadow work.

15 The base color was Scale Color deep red (SC-35). I then added blood red (SC-37) to the base, then continued adding blood red until it was pure blood red. Next, I added Antares red (SC-38) to the blood red, then I gradually added Aldebran red (SC-39).

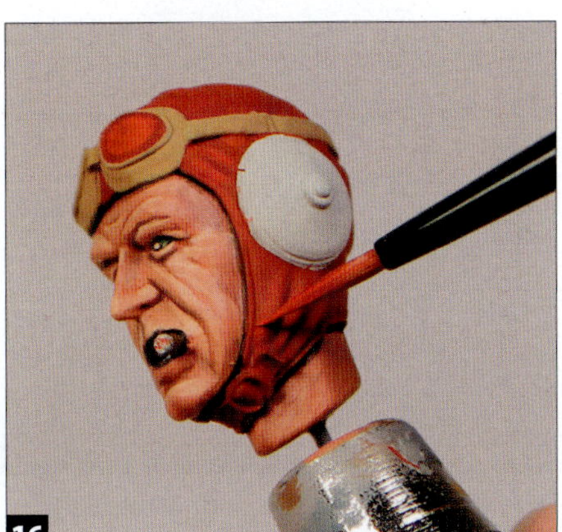

16 The deepest shadow was the base with sunset purple (SC-33).

17 The head is complete with the exception of adding the gloss to the goggles and the metal work to the goggles and buckle.

18 I placed the finished head onto the body to see how it was looking so far. I base-coated the vest at this point too.

19 I like to add the head and different parts to visualize what I am trying to achieve, and it helps to give me a true look at the figure.

20 I base-coated both the life vest and the upper and lower torso uniform. I continued to place the completed head on the figure to see how the colors looked to me.

21 The base coat of the life vest was a mixture from the Andrea yellow paint set (ACS-011). I used the base plus the second shadow.

22 The base coat for the uniform was Vallejo Model Color khaki (70.988) mixed with Andrea khaki paint set (ACS-014) base B4.

23 I began adding highlights to the vest. Again, I made the vest darker than normal so when I added the highlights, the base coat would then become my shadows and all that would be needed would be to add a few deeper shadows later.

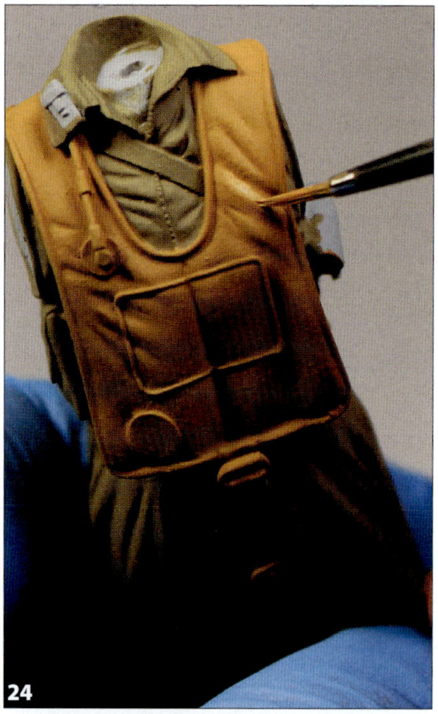

24 I used the base coat plus more Andrea yellow base. The next highlights were just the yellow base. The following highlights were the yellow base plus the first highlight, then I used just first highlight. The highest highlights were first highlight plus the third highlight.

25 The shadows were the original base mix plus the second shadow; for the deepest shadows it was just the second shadow.

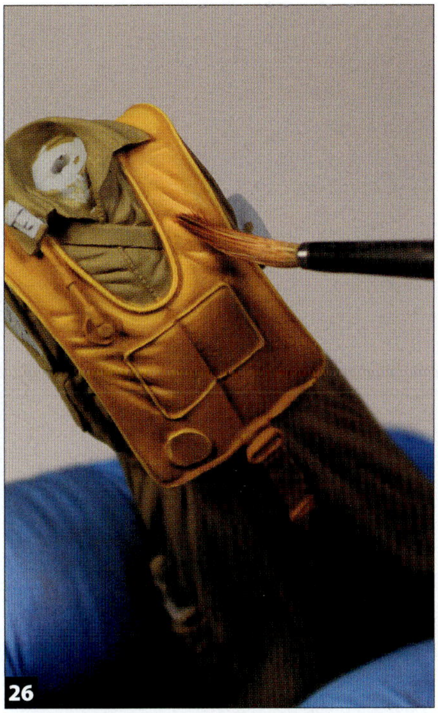

26

When I was done with the vest, the highlights appeared to be too stark, so I used the original base coat and applied a very thin glaze over the areas. This helped to subdue the starkness.

27

I again added the head to the body, this time with the vest completed. I wanted to see if my contrast was the same on the vest as on the helmet.

28

After removing the head, I added the arms to the body and began painting the uniform. The base coat for the uniform was Model Color khaki mixed with Andrea khaki paint set Base B4.

29

Because the areas on the uniform are larger than the other areas on the figure, I was having an issue with paint drying too fast as I was adding highlights or blending colors. To correct this I added Vallejo glaze medium (70.596) to my paint.

30

I added three drops of the glaze medium and mixed it into the paint. This gave me more working time and smoother paint. There were times doing this I had to use a hair dryer to the drying to check how much of a highlight I had or didn't have. Wet paint leaves a false impression.

31

For the highlights I added more of the khaki to the base mix, moving up until it was just the khaki by itself. For the extreme highlights, I added Andrea khaki light A (2), and continued until it was almost pure Andrea khaki light A (2).

The shadows were the base mix plus Andrea khaki paint set base 4, then for the deepest shadows, just base 4.

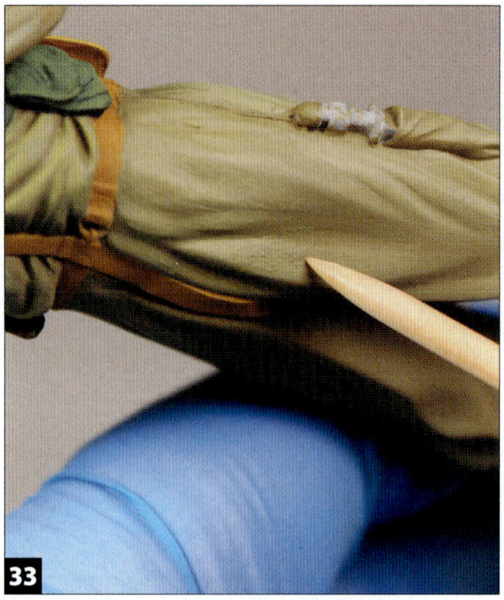

While I was painting, I noticed a few blemishes that somehow I missed during the clean up, priming, and base-coating.

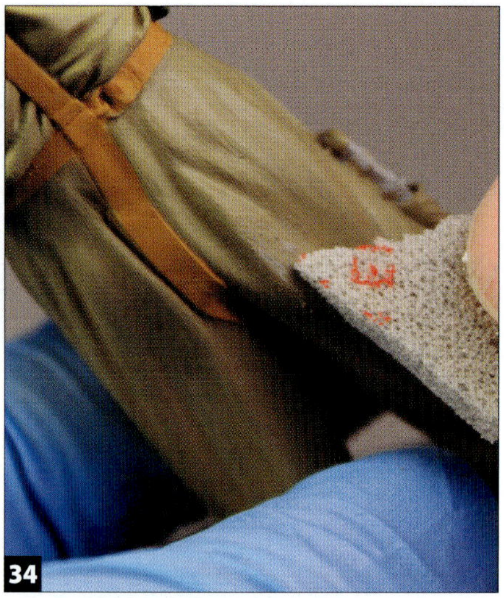

I used 3M fine sponge sandpaper to remove the spots I had missed.

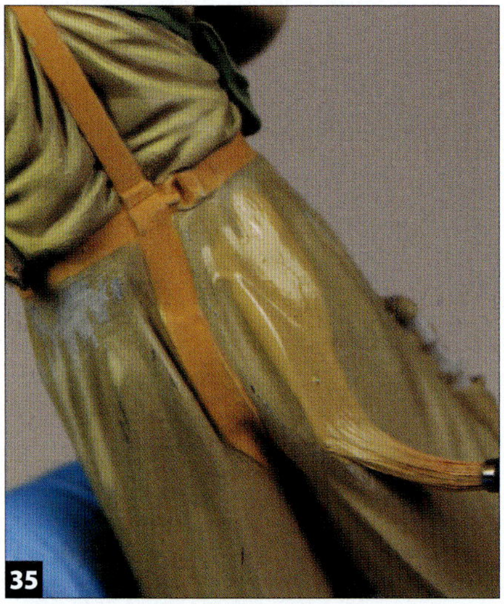

Once I was satisfied I had the areas cleaned and smooth, I repeated the painting process in these areas. Things like this are why I keep my paint mixes simple. It allows me to reproduce them even if I need just a little bit of paint.

I usually have no more than two or three colors per mix. The way I do this is easy. I put a portion of the primary color on a palette. Imagine this color as an eyeball. The next color will be the pupil. Mix them together.

If you desire a third color, add this to the first mix as another "pupil." By doing it this way, I find it easier to control the proportions.

38

I started with a square base and added strips of balsa wood for planking. I attached the wood using superglue and deliberately left some gaps in some areas to make it look more worn .

39

I drilled holes for the pins that were inserted into the feet of the figure to secure it to the base. The next step was to check to see if the figure fit the base correctly.

40

I sprayed the base with Tamiya flat black (XF-1) with my airbrush. I also sprayed random sections on the wood black as pre-shading.

41

I sprayed Vallejo gray blue No. 70.943 over the black. I chose this color because while doing my research on the deck colors of aircraft carriers during this time period I found that they used 250N blue flight deck stain.

42

The final step was to add a coat of Vallejo model wash dark gray (76.517) and black (76.518).

43

I wanted to see if everything looked correct. The metal work was done using Vallejo oily steel (865) and the gloss for the goggles was created with 5-minute epoxy.

44

Obviously a veteran of more than a couple missions, this pilot stands ready to take to the air again.

7 Lightning strikes

Tamiya's new 1/48 scale P-38 builds well right from the box. Painted roundels aided in the weathering plan

ALBERT TURECZEK

The Japanese called the P-38 two planes around one man. The Germans had another name for it: The Forktailed Devil. Certainly, the appearance of the Lockheed P-38 Lightning tipped the balance in favor of the Allies in the air war over the Pacific. This aircraft could fly faster, higher, and had a longer range than its enemies. One mission, however, secured a place in history for the P-38. It was the downing of the aircraft carrying the supreme head of the Imperial Japanese navy and the planner of the Pearl Harbor attack, Admiral Isoroku Yamamoto.

A flight of 16 P-38s from the 13th Air Force, 339th Fighter Squadron, 347th Fighter Group took off April 18, 1943, from a base on Guadalcanal. The

Albert's figures add scale and drama to his well-weathered Lightning.

P-38s flew hundreds of miles skimming 30 feet above the sea to then intercept and shoot down the "Betty" bomber carrying the admiral. It was a tactical and strategic masterpiece, and many historians believe Mission Vengeance was decisive and shortened the war in the Pacific considerably.

There has been some confusion about who downed the plane with the admiral on board, but it is believed it was Lt. Rex Barber in the White 147, *Miss Virginia*. The aircraft was not Lt. Barber's normal plane. It was a loaner because Barber's plane, *Diablo*, was receiving repairs for battle damage at the time. He did not treat the borrowed plane nicely either as he crash-landed the aircraft on return to Guadalcanal, the airframe riddled with more than 100 bullet holes.

1 The parts are cleverly designed so the joints fall on panel lines. Note the leading edge of the wing and top of the cockpit nacelle where separate parts and access panels are located and the actual joints are mostly moved to the sides

2 The build starts with the cockpit, which is crammed with details. I painted the interior Mr Color interior green (27).

3 The instrumentation is included as a decal. There are two panels because there were differences between the F and G variants, which means you have to decide which version to build early on.

4 The P-38 had a steering wheel rather than a control stick. The instrument panel, control column, and radio compartment are in place.

5 With the cockpit painted, weathered, and installed I could turn my attention toward the booms. They have to be assembled separately before they can be joined with the airframe.

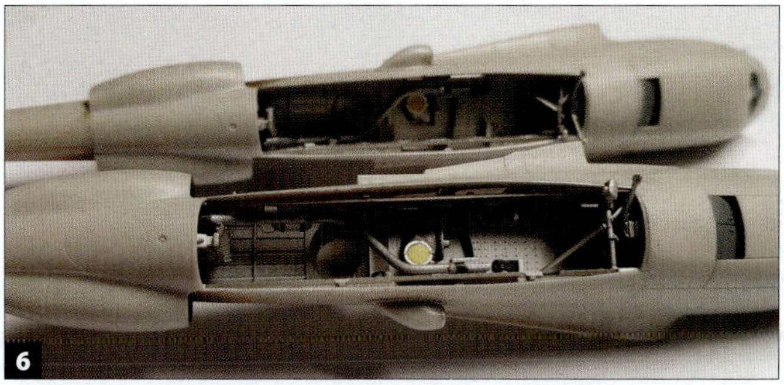

The wheel wells have to be assembled as well before joining the rest of the airfame. Note the space between airframe and wheel well where the landing gear doors can be clicked into place without the aid of glue if you wish.

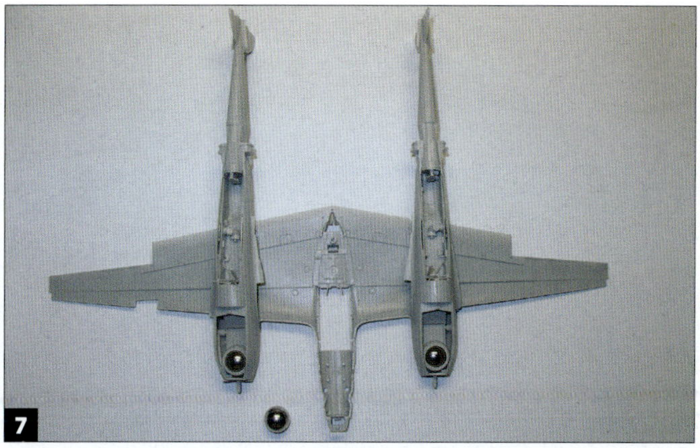

The landing gear geometry of the P-38 would produce a solid tail sitter if you do not counter the heavy tail with ballast. Tamiya has taken care of this in the shape of three steel balls supplied with the kit. There is a socket for them in each boom and just behind the nose. I secured them with a drop of white glue to avoid any rattling later.

The complete tailplane section can be clicked into place. The rudders can easily be removed again, which is particularly useful during painting

This image shows how great the canopy fits. Also note the thin parts, which made me worry about cracks or a fracture on installation.

The kit-supplied pilot in place for a dry-fit. It is not up to the standard of the rest of the kit, so I omitted it. The figure can be removed, and, if you use the decal for the harness, you can just put the pilot into the seat after the model is completed. I found a suitable pilot figure from Reedoak.com. They are 3D printed and incredibly detailed.

The raw build ready for painting. The components literally clicked into place, almost without the use of glue. Note the asymmetric drop tanks as used on the Yamamoto mission, consisting of a 150-gallon and a 300-gallon tank.

Before I could apply paint, I masked the windshield and canopy. Tamiya provides a masking sheet, which makes things easier. The curves of the canopy on the P-38 make single-piece masks for each section impossible. Fortunately, Tamiya provides separate, numbered sections.

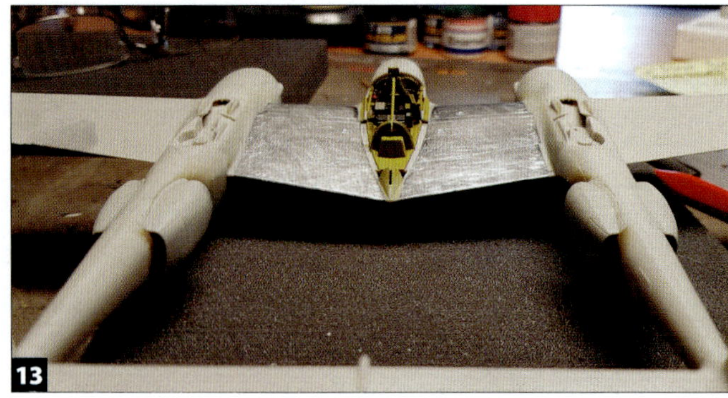

Pilots reached the Lightning's cockpit from a ladder at the rear of the wing, leading to a lot of wear on the wing roots, which served as the walkways for crew and maintenance personnel. In preparation for paint chipping, I covered the surface with Bare-Metal Foil rather than painting the effects on later. This way real metal will show through after I wear and sand the paint.

I masked the painted and installed landing lights with Humbrol Maskol before painting the underside Gunze Mr. Color light gull gray (11).

I airbrushed the first coat on the top with Mr. Color olive drab (12). I decided to forego a primer coat.

The canopy and fairings had to be masked and painted from the inside as well. I used the tight fit of the parts to my advantage here as I could pop them on and off as necessary.

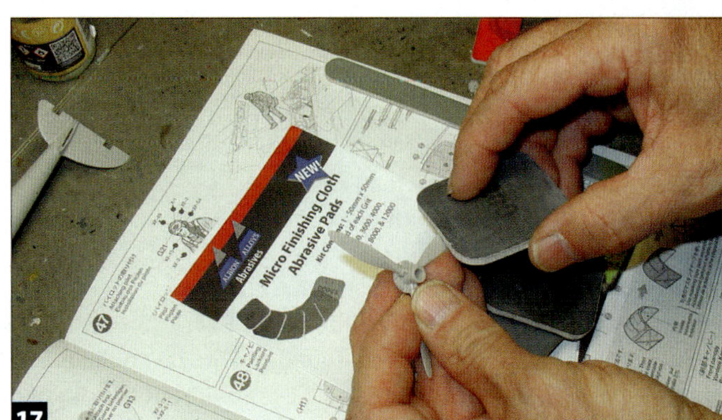

The propellers and spinners comprise an unusually large number of parts, but the fit is excellent and only slight preparation before painting was required.

I painted the national insignia on the plane after seeing the illustration in *Pacific Fighter Aircraft, Profile Book No. 9*, from Swedish aviation artist Claes Sundin, which showed a subtle wear on the insignia. I started preparing by masking the roundels on the wings and booms.

I made roundel stencils from tape using a circle cutter. The first layer of the roundels was painted with Mr. Color flat white (62).

After that I cut the stars and applied them in the middle of the roundels. The U.S. national insignias from this period and location lacked the bars.

Then I airbrushed Mr. Color insignia blue (328) over the star masks. This of course involved a lot more masking and considerable extra work, than using the supplied—and excellent—decals, but I think it was worthwhile for the effect I was after.

This is the painted model with the insignia prior to weathering. The borders between the top and belly colors needed to be soft, and to achieve that I used Uhu Patafix, which is a flexible modeling material, and masking tape.

Weathering would be crucial as it would make or break the realism. This is the case with most models, but seemed more important with this rather simple camouflage.

24

The weathering started with the propellers and had to be subtle. I painted the yellow tips by hand rather than using the kit supplied decals.

25

On the airframe, I post-shaded panels with a lighter tone of the base color. Next I applied a wash of raw umber overall. After this I mottled the surface with darker patches using an airbrush texture mask from Artool. This was done very gently so you can only sense it when closely examining the surface.

26

Now it was time to gently and carefully sand the upper wing surface and reveal the Bare-Metal Foil. In between steps, I covered the surface with a coat of satin varnish before placing the decals and stenciling. After that came several layers of dark wash, before I accented repair patches and panel lines with pastels and weathering pigments from Mr Hobby.

27

The final addition to the model was an antenna made from flexible Lycra fiber.

28

I found some 1/48 scale U.S. Navy flyers, which would lend scale to my model and add a human touch. I know they should be U.S. Army Air Force pilots, but given that Mission Vengeance was flown almost entirely over water, I think they can pass in Navy gear (although a friend already pointed out to me that Air Force pilots would be wearing black shoes). After painting the crew, I weathered them with pastels to match the dusty look of their aircraft.

While it is nothing new to have instruments and harnesses on the decal sheet, Tamiya re-created two signature features of the P-38 with decals. One is the cooling holes on the prominent gun barrels. This worked so well the brass set I bought for the model stayed in the drawer. Try to see if you can spot it being decals!

The other feature is the mirrors on the inside of each engine nacelle and the rear view mirror. The oval-shaped mirrors on the cowls allowed the pilot to visually check the gear. They were thin metal foil on the decal sheet, which had to be sealed to stay in place. Less successful was the decals for the shiny oleo struts. I could not bend them properly around the landing gear strut even with a strong decal solution. I replaced them with Bare-Metal Foil Super Chrome.

With that, *Miss Virginia* was ready to go.

8 Getting his feet wet with IJN *Tone*

Tamiya's 1/350 scale model of the cruiser proved to be a perfect introduction to ship modeling for an airplane guy

CHUCK DAVIS

I've always been fascinated by Japanese ship designs from World War II—they manage to look purposeful, yet elegant. Much thought was put into designing for a purpose, and none can be a more obvious example than the "Scout Cruiser" typified by the *Tone*. Designed and equipped from the outset with the ability to handle scouting aircraft, and famous as the cruiser that had a catapult failure in the Battle of Midway that delayed the discovery of the American fleet, the *Tone* fought across the Pacific and the Aleutians until the end of the war. I've always wanted to build a Japanese cruiser, so what better way for an airplane modeler to take on ship modeling than building the *Tone*?

Tamiya's 1/350 kit of the *Tone* includes gorgeous detail and some photo-etched (PE) metal parts, but naturally I wanted to add some more brass to challenge myself—or I got caught up in the moment, maybe.

The Tone *offered Chuck a chance to apply his skills as an aircraft modeler to a new subject. The result was an award-winning model.*

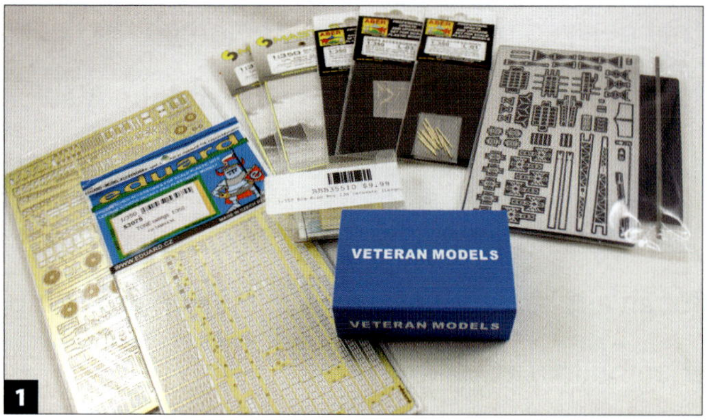

1 Tamiya's *Tone* has a ton of nice detail, including some PE parts—but especially with ship kits, there's always more you can add. I stocked up on brass barrels for the main and antiaircraft weapons from Aber and Master, exquisite resin searchlights from Veteran Models, paravanes from Big Blue Boy, and of course—PE details and railings from Eduard.

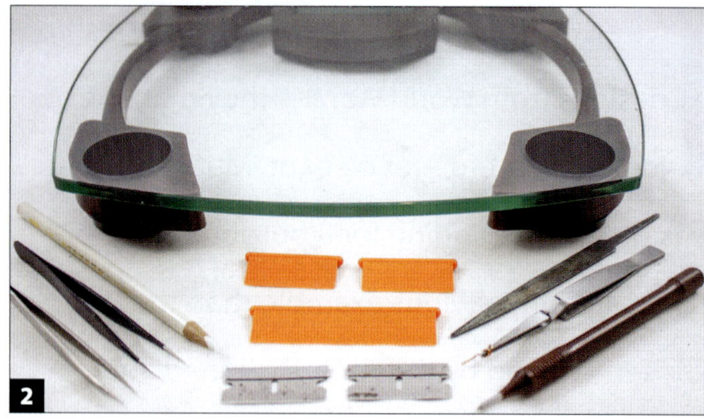

2 With all the heavy metal bending involved in this kit, I pulled together all my tools. I use an old heavy-duty, glass-topped bathroom scale to give a solid surface for cutting and bending. Files, tweezers, glue applicator, scalpel, and "part picker" are all important tools. Razor blades and other specialized bending aides are also ready. And of course—good lighting and vision assistance are mandatory.

3 I use a standard No. 11 blade in a hobby knife to remove parts from brass frets, cutting well away from the part. A straight razor blade makes short work of cleaning off the remaining brass nibs.

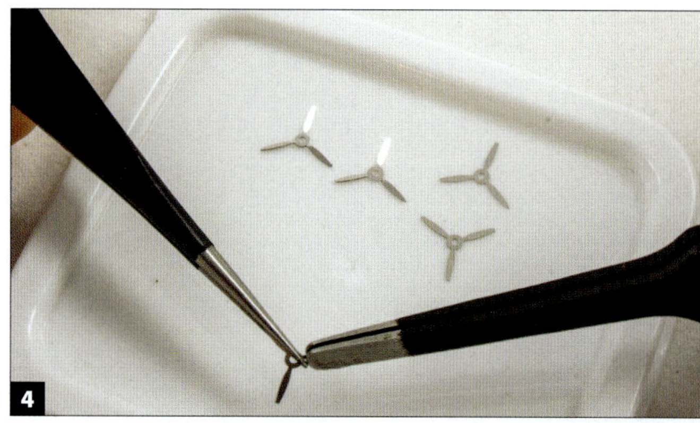

4 So, where would an aircraft modeler start his first ship kit? With the aircraft, of course! Even though it carries 8-inch guns, the *Tone*'s actual "main battery" is a fleet of five Aichi "Jake" reconnaissance floatplanes for its role as a scout cruiser. The Jakes in Tamiya's kit are quite nice and include PE propellers—I added a twist to each blade to enhance realism.

5 In the model aircraft world, Tamiya is known to have marginal decals. Apparently the same is true for its ship models—the canopy decal for the Jake would not conform at the front or rear of the canopy.

6 After numerous attempts with setting solution, I simply cut the stubborn parts away and painted them instead. You can see much work is still needed to avoid decal silvering.

7

After finishing five airplane models, I still had a ship to build …

8

Tamiya's hull design includes several support bulkheads that can be modified for a waterline display. The fit was precise, but I added numerous clamps just to make sure there were no problems surfaced later.

9

Before proceeding much further, I wanted to nail down the stand for the *Tone*. I found custom-machined brass stands online. A trip to the craft store for a simple wood base was followed by a trip to the home improvement store for hardware and pads.

10

I measured and marked off the center and centerline of the base, followed by marking the attachment points Tamiya molded into the hull for its plastic stand.

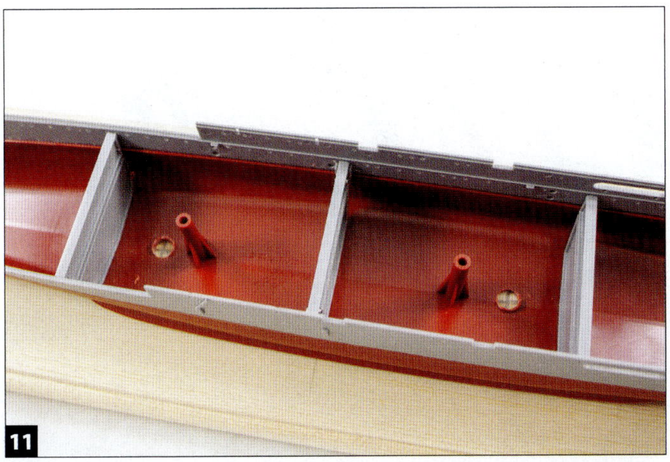

11

I drilled out the mounts in the hull and double checked the spacing on the base.

12

My dad taught me a simple woodworking trick—use tape to keep wood from splitting when you drill though the opposite side. I put tape on the "good" side where the holes will be.

13 I drilled pilot holes through the marked attachment points with a drill bit sized to allow the hardware to pass through cleanly. I used a larger hole drill bit to countersink the bolt head and washer below the surface of the base.

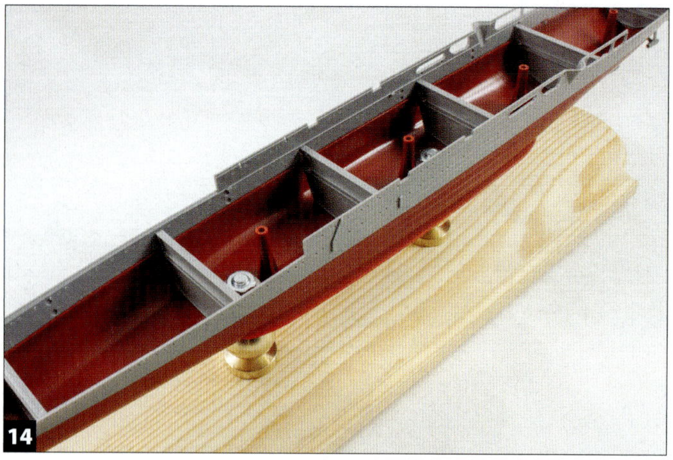

14 A quick test-fit showed all was well. I stained and sealed the base, then attached the unpainted hull to it, making sure to use locking nuts on the bolts—I don't want the *Tone* to suddenly un-dock itself!

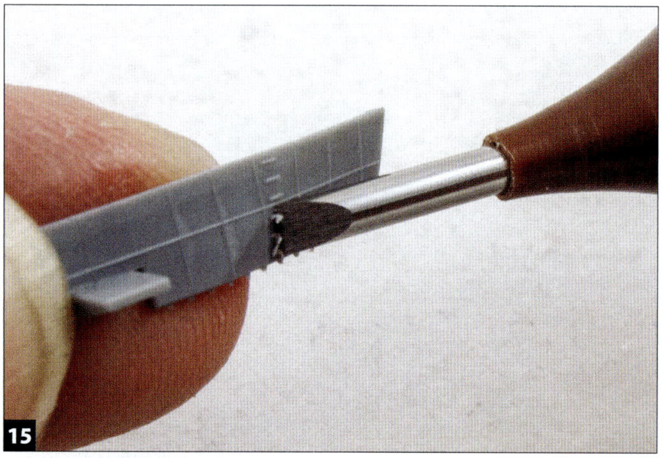

15 Regardless of the model I'm building, I spend a lot of time up front preparing parts. Many of the tiny details molded in place by Tamiya will be replaced by PE, requiring removal. A micro chisel is a handy tool for this task.

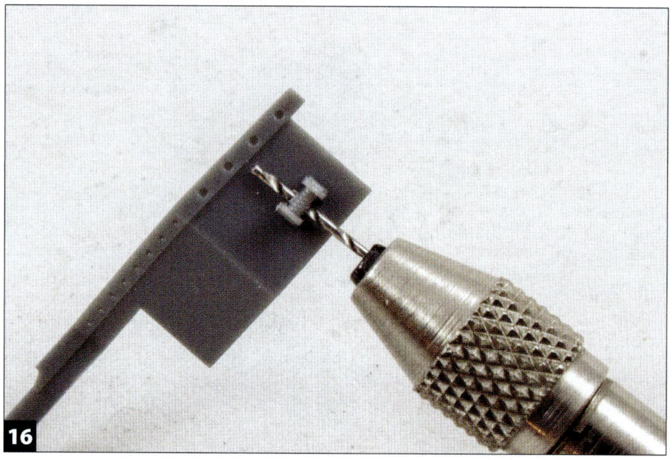

16 I looked for any other simple ways to improve the model and realized I could use a drill to open up a cable reel molded on one of the superstructure walls.

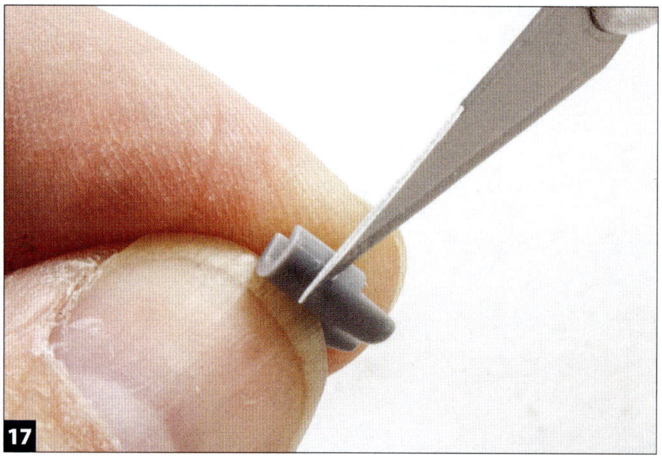

17 Even though the scale is small, you still need to remove visible seam lines.

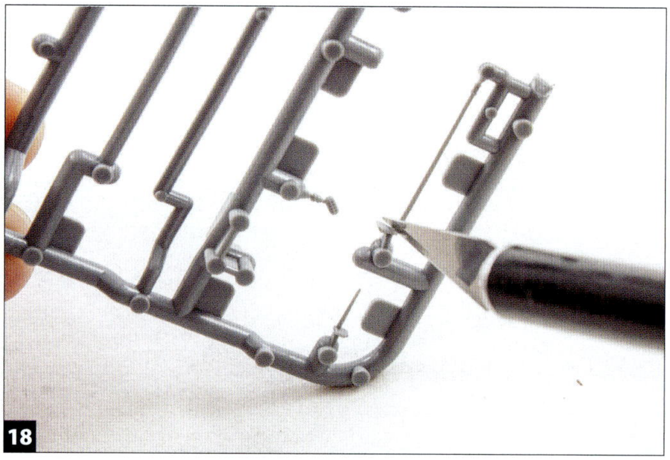

18 Some preparation work is better done on the tree to avoid breaking fragile parts.

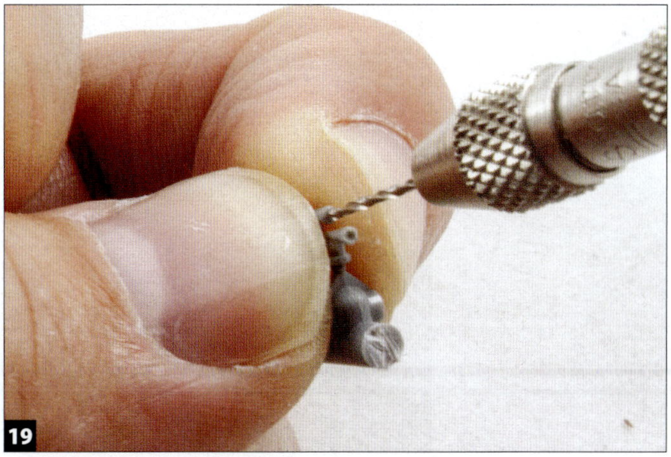

I spotted some cone shaped parts that I thought were sirens. Careful drilling gave them a realistic opening, an easy way to add a bit more detail.

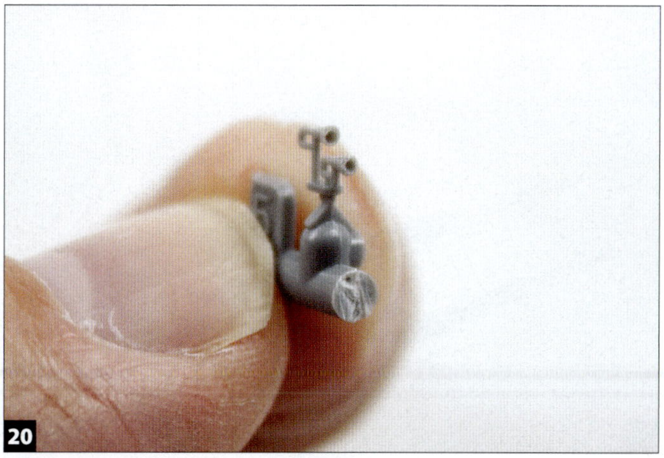

I found out later that they are radar antennae—but they are still cone shaped.

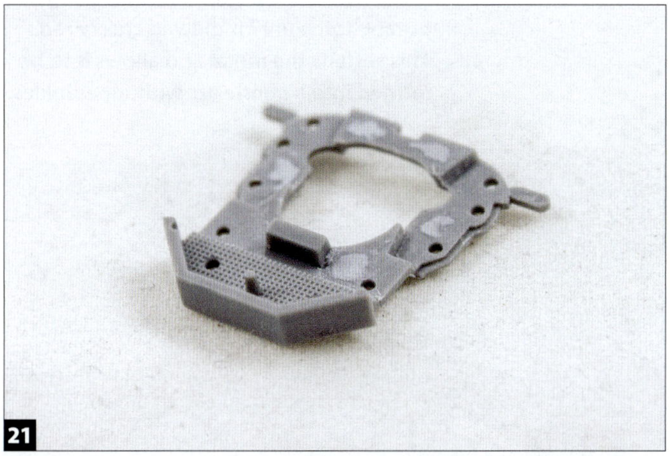

Even with Tamiya's expert molding, you still must look for and address surface defects like the ejector-pin marks on the lookout platform.

I've never been a big fan of PE details, but they're a necessary evil with ship models. Most bends were made with simple straight razor blades or tweezers. I trained myself by bending ladders—after a couple of failures. The side rails are bent most of the way to vertical—but not fully, otherwise the rungs can't be folded. Don't ask how I found that out …

Using the tips of the tweezers, the rungs are gently pushed toward their folded position.

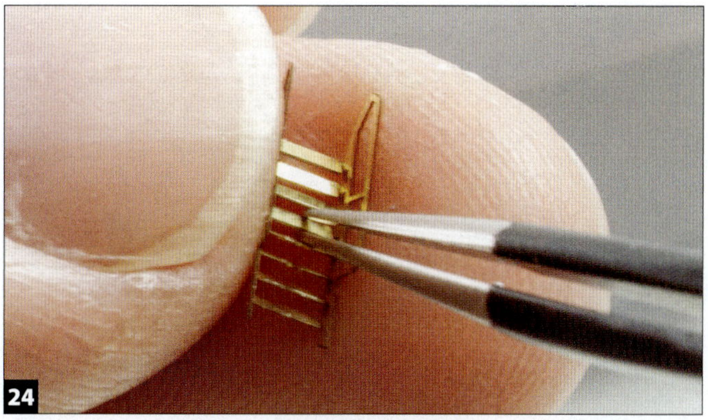

The tweezers can then grab the rungs, and a gentle squeeze pops the rungs in place.

25 A further gentle squeeze on the sides completes the vertical bend of the side rails.

26 Details like ladders with side rails and open rungs are what makes a ship model stand out.

27 Some PE details need to be circular. Smaller pieces can be wrapped around a gauge pin or other suitable round object. Larger details like this funnel shroud need a bit more effort. I heated the shroud using a butane torch until it glowed cherry red. This softens the metal and allows it to be formed into a gentle arc without wrinkles.

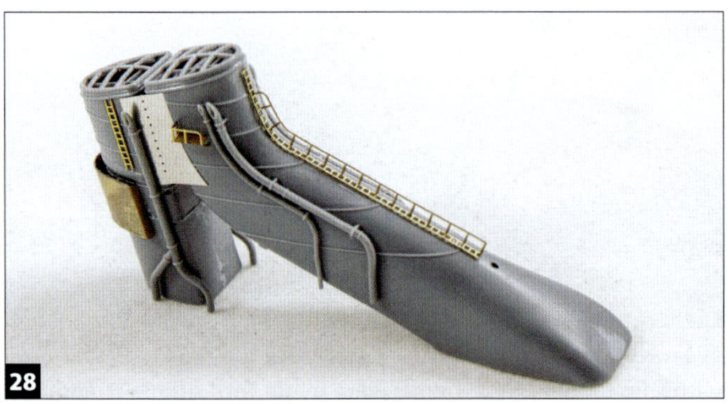

28 Wrapped cleanly around the rear of the funnel, you can see that without the annealing process described, the part would have wrinkled or bent instead of producing a gentle arc. The silver PE parts are supplied by Tamiya in the kit.

29 Another very handy tool is a "part picker pencil" with a sticky tip. They make precise placement of parts like this ladder simple.

30 A fully assembled turret shows the added details—new brass gun barrels, access hatches, and ladders complete the assembly.

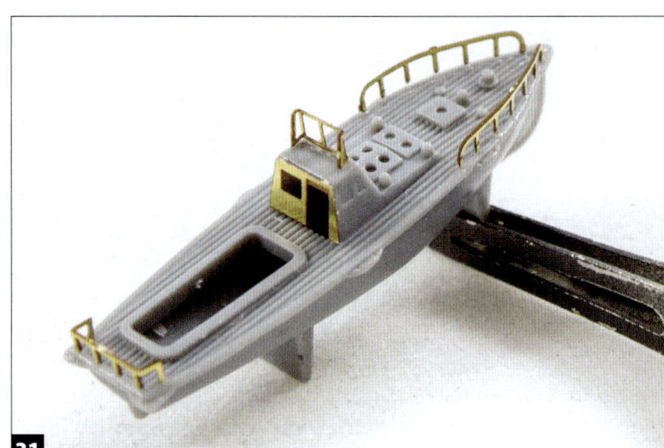

31 Even the ship's boats get more detail. Placing the PE before painting allowed me to ensure the parts were firmly attached.

32 The aft mast is a complex assembly. A host of metal parts were added—including the crane rigging supplied in the kit.

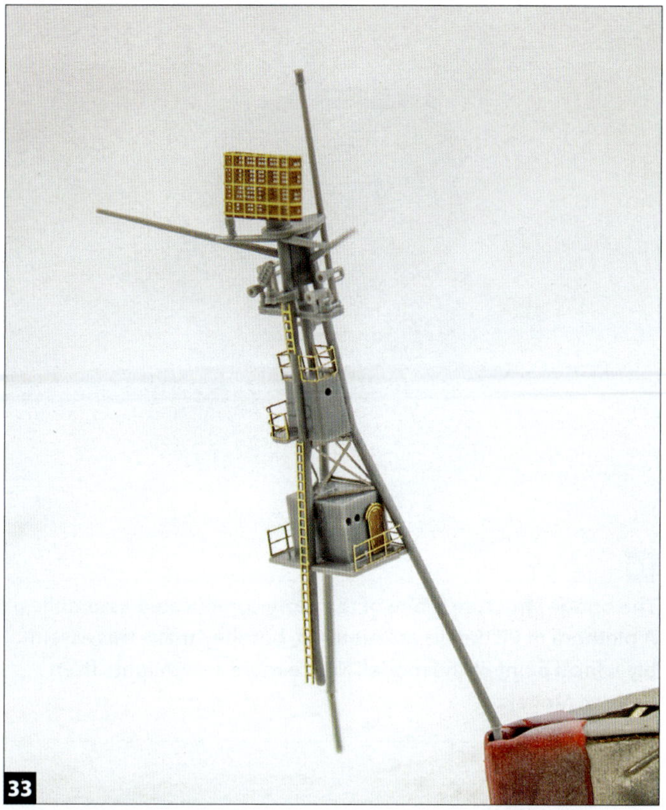

33 The forward, main mast is highly detailed as well. Eduard includes the main radar antenna, which folds easily despite its delicate appearance. Notice I've drilled out all the portholes. I did the same to the hull.

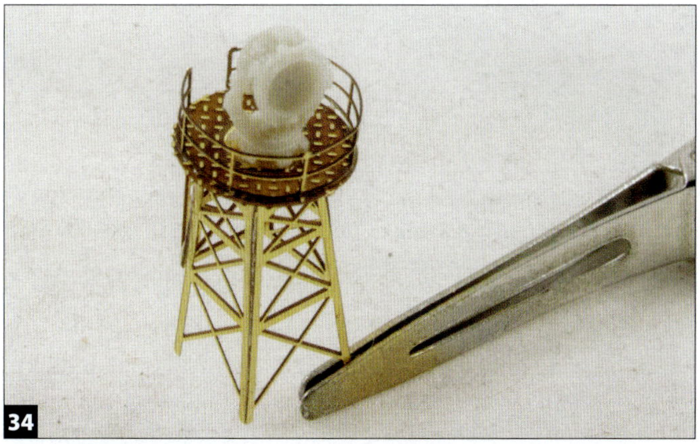

34 One detail that almost made me jump ship—the searchlight platforms are completely replaced by PE parts. They are exquisite, but frighteningly fragile. They also ended up very difficult to mount in place on deck because the legs are so thin. The resin searchlights from Veteran Models are beautiful! Painted with Model Master Metalizer steel with a Tamiya smoke wash, they look real.

35 Most deck structures had at least a handful of railings and details to add. Careful study of the instructions and test fits are a good idea to avoid problems. I didn't get the placement of the tiny ladders on this deckhouse exactly right, causing fit problems when I tried to install it to the deck—lesson learned.

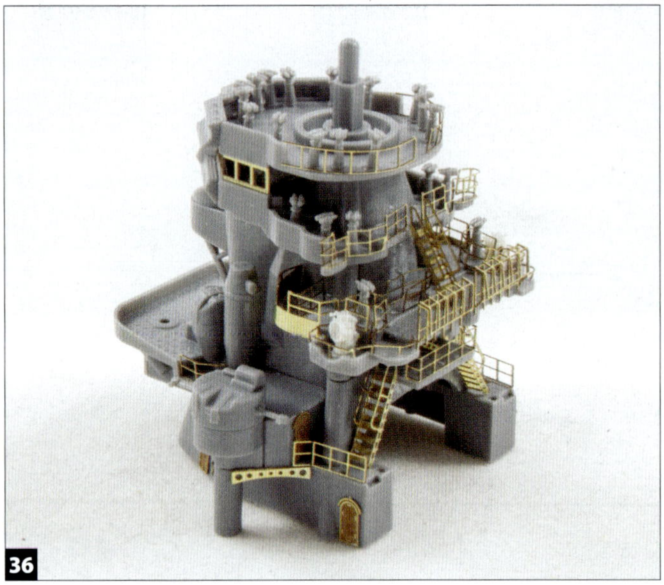

36

The bridge structure is one of the more complicated assemblies. A plethora of PE details are required, but they make this assembly a focal point of the model. Notice more searchlights from Veteran Models.

37

As I was checking the fit of the main deck and arguing with myself about methods for applying the railings, I realized the deck pieces correspond to the linoleum portions only—the deck edges are the hull sides and supporting walls. That gave me the idea of attaching the railings to the hull sides and wall tops, instead of attaching them after painting. That would allow them to be firmly attached before being painted along with the rest of the hull. The aft support wall for the upper deck illustrates what I mean—all railings are already in place.

38

Now that some of the preliminary assembly work was underway, it was time to move the hull forward. Having previously sprayed the lower hull with Tamiya dull red (TS-33) and sealed it with clear flat, I masked the hull and the attached base to prevent overspray.

39

The deck was airbrushed Tamiya linoleum deck brown (XF-79). I added shading using a mix of Tamiya brown/JGSDF (XF-72) and deck tan (XF-55).

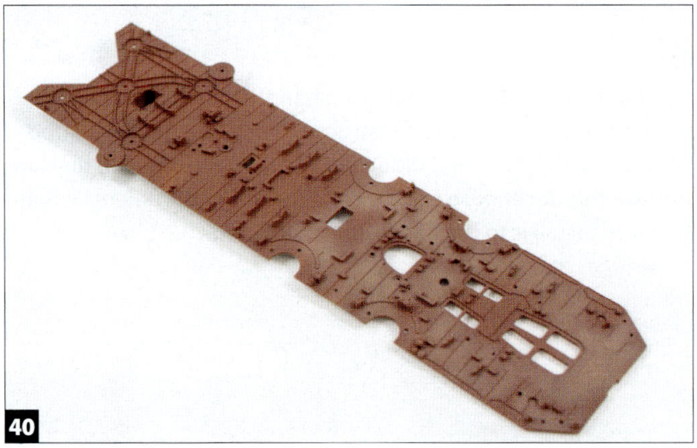

40

A close up of the upper deck shows the fading in process. This breaks up the monotony of a solid expanse of color.

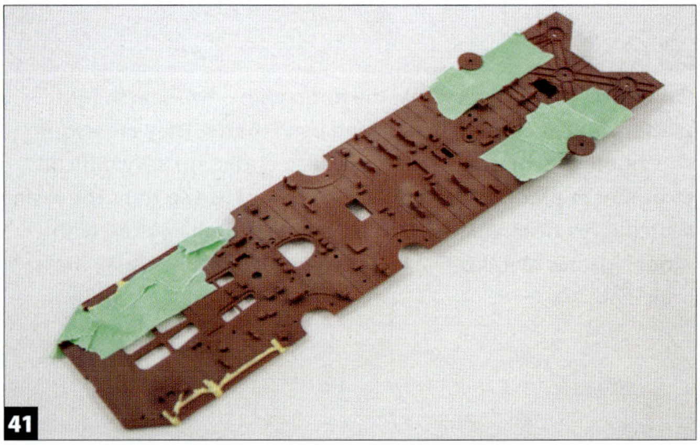

41

Once sealed with an acrylic flat coat, I masked off the major areas that were gray—I didn't bother with masking the tiny deck fittings, since I actually wanted to complete the kit.

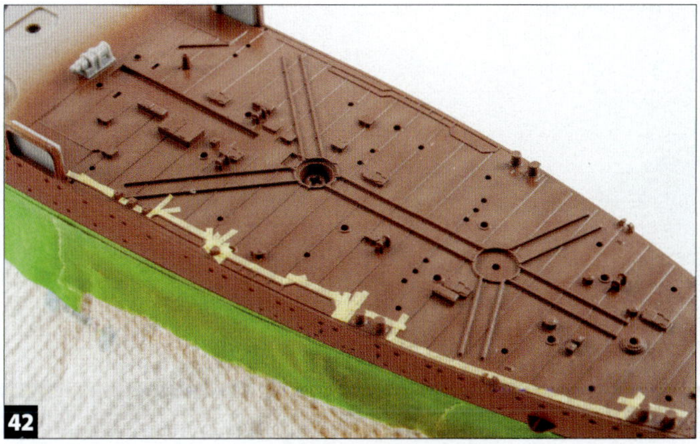

Tiny strips of thin masking tape outline the gray areas. Notice the railings are not in place during masking to avoid damage, but the deck edges where they will mount will be painted gray.

After covering the remainder of the deck area, the railings are installed just prior to painting the gray.

The hull, deck edges, and railings were primed using Alclad gray Microfiller Primer, and then painted with Tamiya IJN gray (Sasebo Arsenal) (XF-77). Fading was accomplished with a streaky application of a lightened mix of the IJN gray.

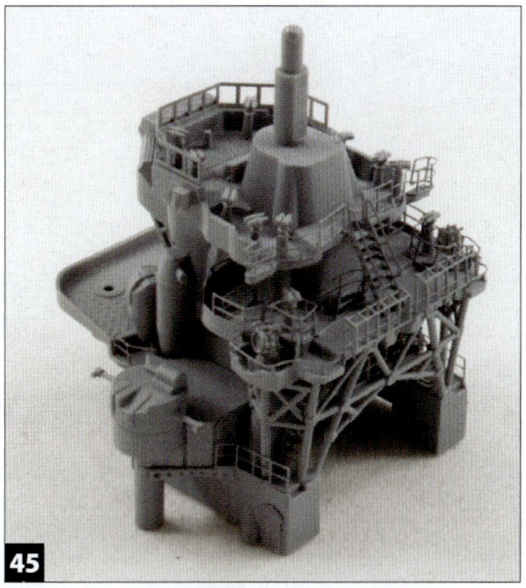

First step in finishing the bridge assembly was a good priming, again using the Alclad gray primer.

Next was a coat of the Tamiya IJN gray. Make sure to spray light coats from all angles.

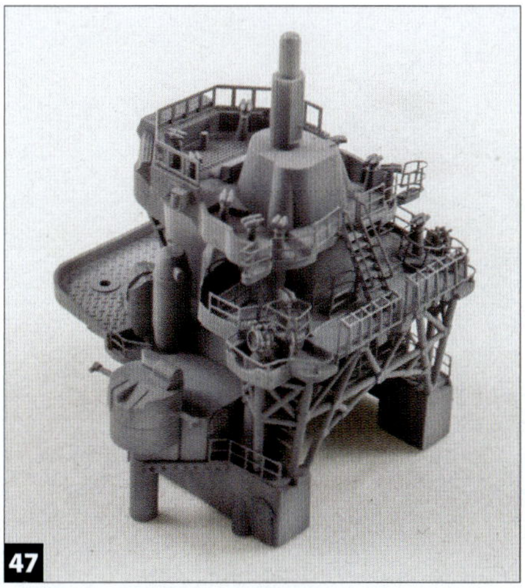

Gentle fading with a lightened mix of the IJN gray lifts detail out of the shadows. An acrylic clear gloss seals everything for the upcoming weathering steps.

Now for something different—and a bit scary. I applied a dot filter using oil paint. I only used white for fading and brown for rust. I applied the dots sparingly, so as not to overwhelm the model.

Once blended, the oils give a soft weathered appearance to everything. Notice the faint rust stains at the base of the bridge assembly.

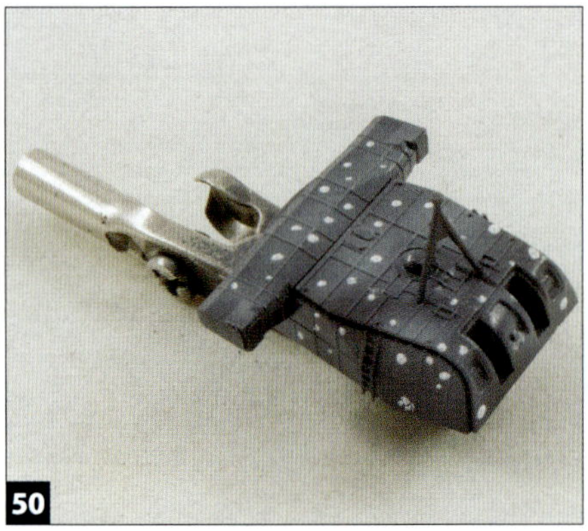

50 The same process was performed on all major deck assemblies, including the main turrets. I confined the dots to white for them.

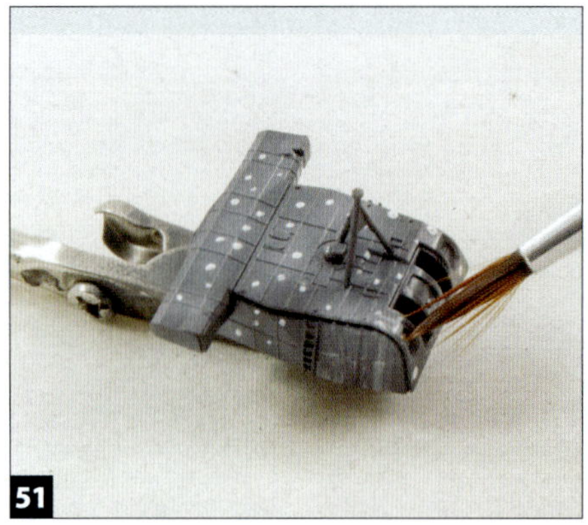

51 I dampened a brush with Turpenoid—a synthetic turpentine substitute—and wiped almost all of it off. Using a nearly dry brush, the oil dots are blended and "pulled" into place.

52 Some of the oil paint sludged into the corners, especially the aggressive panel lines on the turrets. Wherever the fading appeared too aggressive, I simply misted a thin coat of the base gray over the top to knock down the contrast.

53 After another coat of clear gloss to seal the fading, I used Flory Models dark dirt wash to add more weathering. It is simply flowed on with a brush …

54 … and then wiped off with a cotton swab or other soft tool.

55 The same treatment was applied to the bridge—seen here in the midst of the wash process.

56 Simple details matter on these ships—Eduard includes PE pieces for the navigation lights mounted to the bridge. The brass housing has a small chunk of plastic rod inside to simulate the light. This was painted Testors Metalizer steel, then coated with Tamiya clear red (port) and green (starboard).

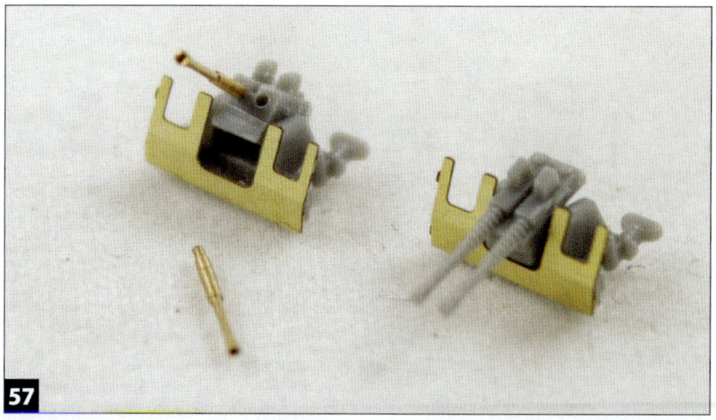

57

Yet more details—the antiaircraft guns have barrels replaced with brass, and Eduard includes the shields in its details. I also wanted to avoid a stiff, "staged" look in general, so I modified all the gun mounts to have varying degrees of elevation by slicing off the bottom of the mounts at random angles.

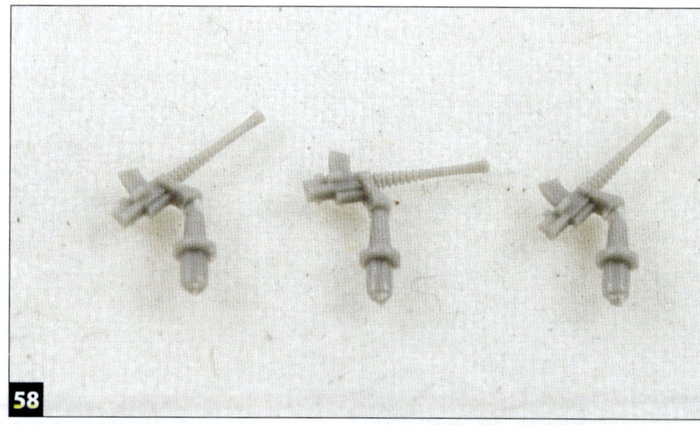

58

I also modified the single 25mm gun mounts for the same reason, by bending the gun mounts at various angles.

59

The kit includes decent replicas of the paravanes carried by the *Tone*, but they can't compare to the brass replacements from Big Blue Boy.

60

The ships boats were all carefully painted and weathered. One disappointing feature is the holes molded into the three whale boats to allow mounting to the boat davits—only two are mounted that way, leaving the third with holes in the hull. They also have knockout marks in the floorboards that are nearly impossible to hide. Maybe next time I'll find replacement whaleboats too!

61

The hull details were slowly assembled—I experimented using Flory Models Mud Brown wash for rust stains on the secondary armament tubs. Not bad, but hard to make into a streak.

62

I applied oil-dot filtering to the hull. Note that I haven't applied any of the brown oil paint below the waterline.

63

I used various soft brushes to blend the white and streak the brown oils, teasing them into place until they looked about right.

64

The brown rust stains were stippled in place being careful to not go too far below the waterline.

65

Flory Models dark dirt wash was applied to the decks, carefully hitting all sides of the tiny deck fittings.

Having rigged many a biplane model, I was not worried about rigging the *Tone*—until I realized I didn't have any references showing where the wires went. The box top was of some help, but I couldn't find any readily available diagrams. I added some obvious wires using Uschi standard size elastic rigging. A drop of superglue at each attachment point is quickly solidified by brushing some accelerator onto the elastic thread.

While I was studying the box top, I noticed Tamiya shows a wing from a Jake floatplane stashed up against the funnel. The instructions show "aircraft spare parts racks" made up from etched details in the kit, so I chopped up a wing and float assembly from one of the spare Jakes, added paint and decals, and mounted them to the racks—a quick way to add a bit of color.

I had been caught up in the moment, but then I realized the model was finished. And with that, the moment was gone, and I was free.

9 A Canadian in the RAF

Tamiya's 1/32 Spitfire makes an excellent starting point for this South East Asia Command fighter

JOHN MAHER

Tamiya captured the Spitfire's wing dihedral perfectly. Brian and I wanted to show F/O Ardeline having just landed from a sortie. Note the blown out patches over the .303 machine guns. Typically, when a fighter lands, the first thing the crew does is service the guns and re-fuel. Jerry cans are at the ready to top up the wing fuel tanks. They also help fill "dead space" around the Spitfire.

Paul Ardeline joined the Royal Canadian Air Force in 1941, and like many Commonwealth pilots, he found himself posted to an RAF squadron. In January 1944, he arrived at 152 Squadron, then based in Burma. Known for its Leaping Panther emblem surrounding the left fuselage roundel, Flying Officer Ardeline further personalized his Spitfire by adding Canada's national symbol, the maple leaf, to the fuselage roundels. It was unusual that he was allowed to so mark his aircraft, as RAF regulations prohibited changes to national insignia.

My friend Brian Wildfong and I collaborated on this joint project to model this Canadian in the RAF and his uniquely marked aircraft with Brian creating a figure and me building Tamiya's 1/32nd scale Spitfire Mk VIII. Brian represented F/O Ardeline with parts from a variety of sources while I used aftermarket parts that either added detail left out by Tamiya or actually improved ease of assembly.

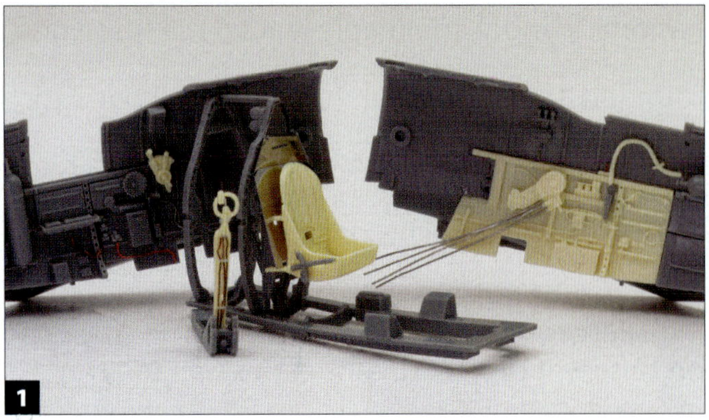

1

2

Beginning with the cockpit, I used the following resin sets from BarracudaCast: seat No. 32001 (which features the seat back pad missing from the Tamiya part), upgrade set No. 32003 (which includes the missing oxygen hose), and starboard side wall No. 32008 (which includes missing wiring details). Everything is a drop fit except for the starboard side wall. It requires you to scrape away existing plastic details for it to fit. I also added wiring using fine copper wire and solder.

Tamiya paint was used exclusively for all airbrush work starting with a 50/50 mix of cockpit green (XF-71) and white (X-2) to create RAF interior green. I picked out details with Vallejo paint, then applied washes using both enamels from Ammo by Mig Jimenez and oil paint from 502 Abteilung.

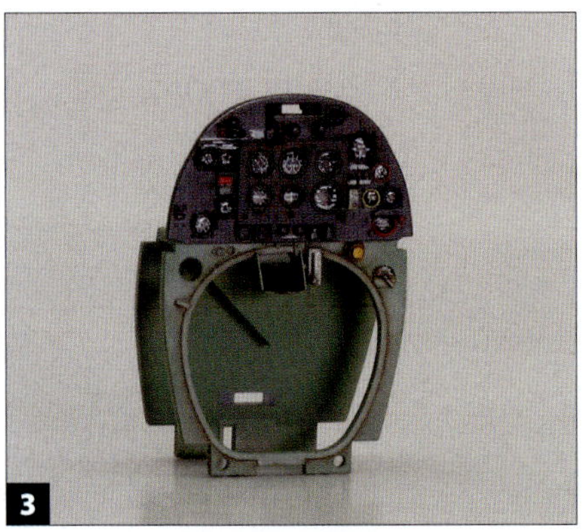

3

Tamiya has you put its instrument decals on the back of the clear parts representing the instrument glass, but I think this distorts the faces. Instead, I punched out each of the dials individually and put them on the front of the clear parts and then sealed them with Tamiya clear (X-22). The panel was first airbrushed with German gray (XF-63), and details were picked out with Vallejo acrylics. A wash of Winsor & Newton lamp black accentuated the details in the kit-supplied photo-etched (PE) metal main panel and the excellent decal placards included in BarracudaCast's upgrade set finished everything off.

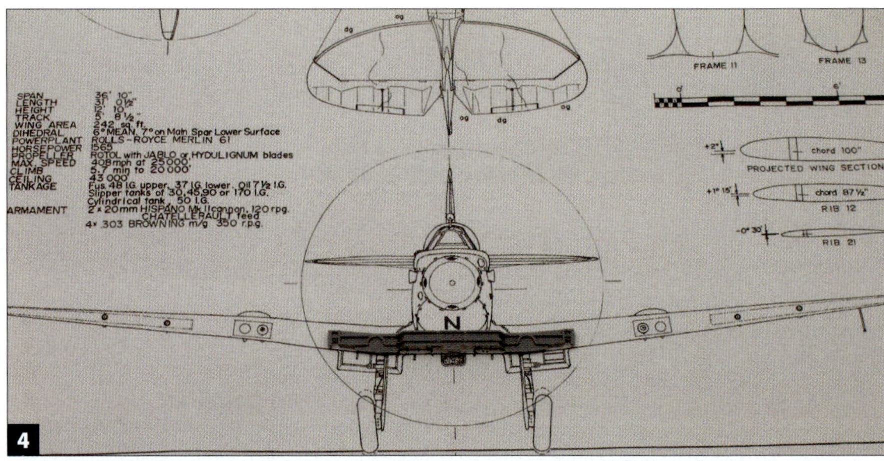

4

When Tamiya first released its big Spitfire series in 2009, there was some discussion suggesting the company had not captured the Spitfire's distinctive dihedral. However, I found the kit's center spar (part A10) lined up perfectly with what are regarded as the best Spitfire drawings by A. R. Clint, so Tamiya in fact has it right.

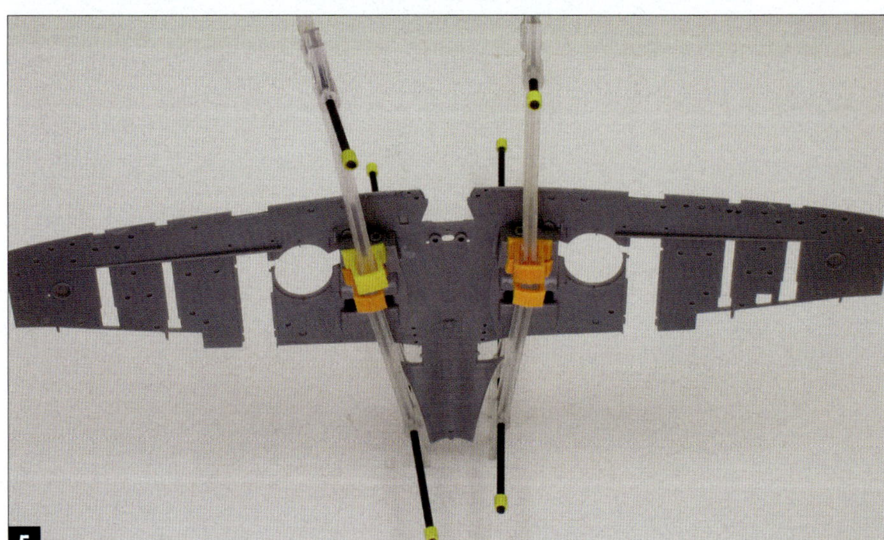

5

I followed a wing assembly sequence that ensures you get nice, perky wings. Begin by securing the inboard sections of the wheel wells (parts G17 and G18) in place with Berna clamps and then flow liquid cement around the outside edges.

6 Then repeat the clamping/liquid cement step with part A10, the center spar. Note that the lower gun panels are not yet fitted and therefore how much open area there is. Resist the urge to install the gun panels and you will be rewarded later.

7 Repeat again with the outer sections of the wheel wells (each is a seven-part assembly). Note how the Berna clamps also suspend this assembly while the glue sets so nothing can throw it out of alignment. It's critical this alignment be right as it ensures you have the correct geometry for the dihedral in the center section of the wing. The wing top sections will be used to ensure correct dihedral out to the wingtips.

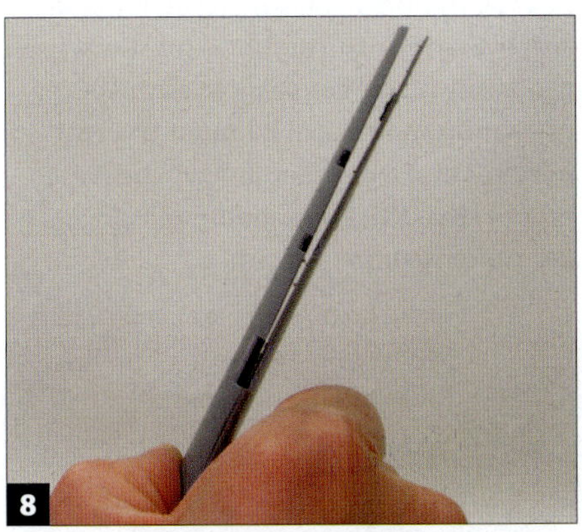

8 Before mating the wing halves, I filed the leading edge of each wing half so that when joined, there is a V-shaped groove. In this dry fit, note how the lower wing naturally droops away from top wing. It is for this reason that you don't want to add the gun panels until the wing is fully assembled—doing so now could prevent the lower wing sections from having enough flex to be pulled up to its upper sections.

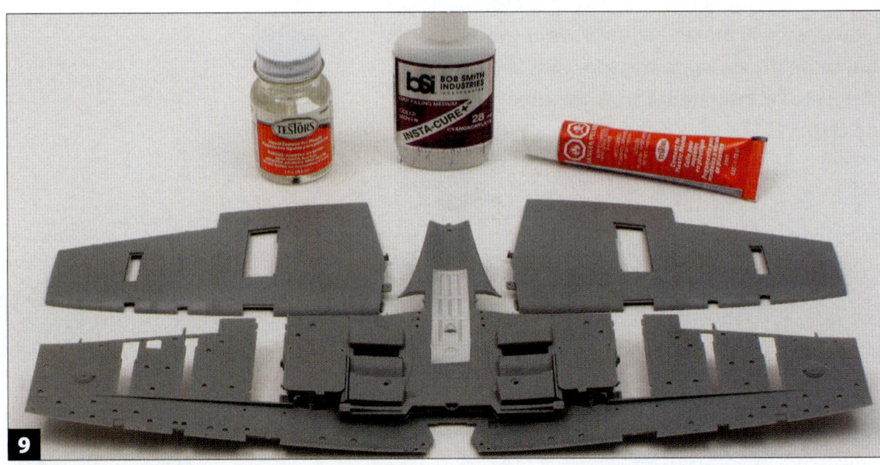

9 You'll need both Testor's liquid and tube cement as well as gap-filling superglue to complete the wing assembly.

10 First step is to draw the bottom wing up to the top wing—you will want a strong joint, so I used Testor's tube cement applied sparingly on the top wing along the molded-on spar and on the area above the wheel wells.

11

Once again, Berna clamps hold the parts in place. Use liquid cement on the trailing edges of the wing. Note that the gun panel covers are still not in place. Tamiya does call this out in its instructions and this is key. Resist the temptation to attach them first by applying glue from behind. If you do, it will make the lower wing too rigid to be drawn up to the top wing sections, which could then throw the dihedral out.

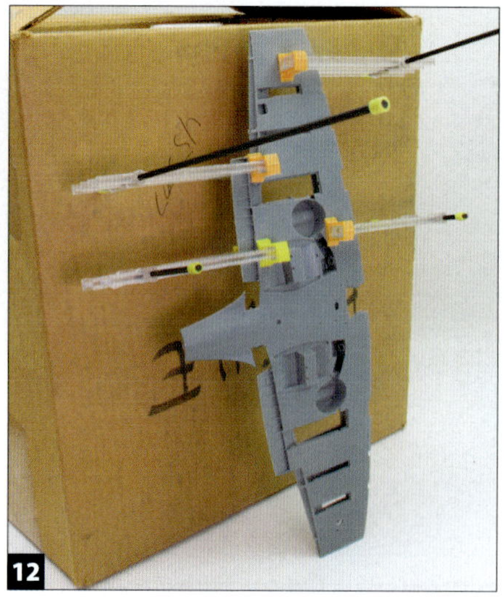

12

Each wing was done separately and then set aside to allow the glue to set. You want to position the wings so nothing can interfere with their alignment.

13

I used gap-filling superglue to seal the wing leading edges. After first running superglue accelerator into the leading edge groove I added earlier, I apply superglue using the eye of a needle that has been ground down to form a U. For the next 30 minutes, the superglue is actually softer than the surrounding plastic, so you can easily blend the seam in with 1000-grit sandpaper. Leave it longer than 30 minutes and the superglue will become harder than the surrounding plastic, making sanding more difficult.

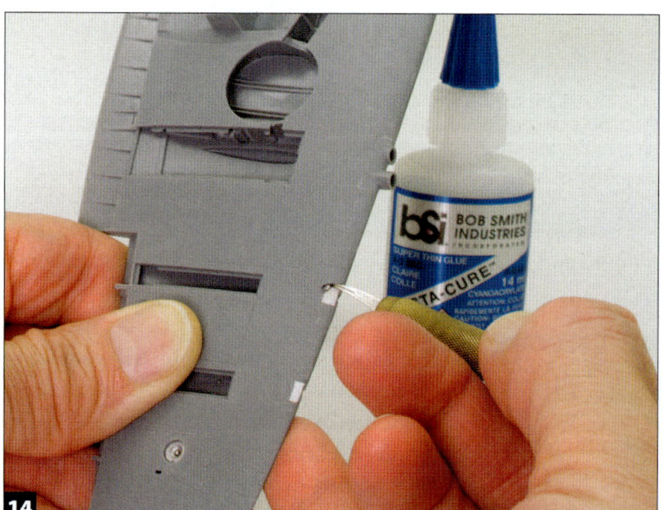

14

The only fit problem I've ever had with a big Tamiya Spitfire kit has been the inserts for the .303 gun ports—they're slightly undersized. Small squares of .0005-inch plastic attached with thin superglue were used to blend in these parts. At this point I have positioned the upper gun panels, for which you can use liquid cement applied from behind through the open lower wing gun panels.

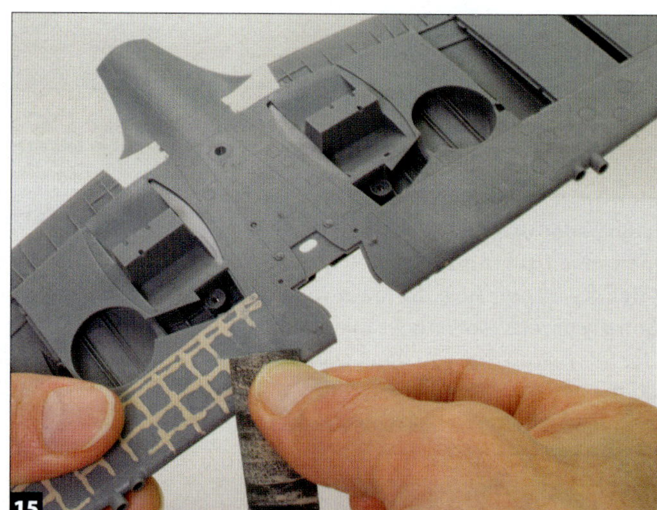

15

Just like the Mustang, Spitfires had sections of their wings puttied at the factory to improve aerodynamics by smoothing out turbulence caused by rivets. As per factory specs, I filled in the rivets from the leading edge back to the main spar using Humbrol No. 121 pale stone matt paint.

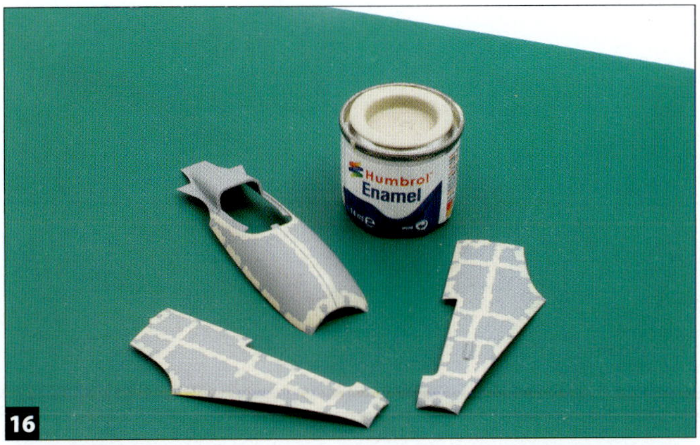

16

I also filled in the rivets on the cowlings and the landing gear leg covers, again using Humbrol pale stone.

17

The last step in building the wings was to finally add the lower wing gun panels and then mate the finished wings to the fuselage. Painting began with the white South East Asia Command (SEAC) stripes on the wings, tail planes, and tail fin.

18

After masking off the white strips, I pre-stained the paint with German gray. I also masked off the areas where glue is needed to add the radiator faces and covers. Toothpicks in the landing gear locators keep the model off the workbench during painting and finishing.

19

SEAC Spitfires were finished in dark green and dark earth camouflage. I used dark green (XF-81) and a mix of khaki (XF-49) (85 percent) and flat earth (XF-52) (15 percent) respectively. I mixed lighter shades of each by adding up to 15 percent white. Paper masks were used to demarcate the two upper surface colors.

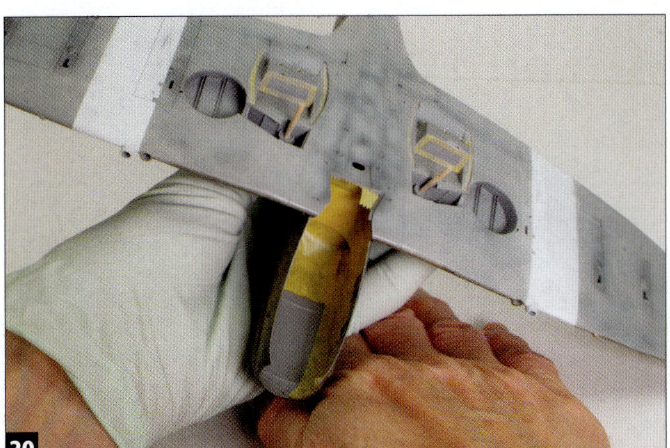

20

The undersurface was finished with medium sea gray (XF-83). Note how the upper cowls are held in place with Tamiya tape so I could paint them separately from the engine assembly.

21

All the markings are painted on with custom masks from Mal Mayfield of Miracle Masks. The leaping panther emblem was created with a series of masks I added and removed, beginning with the white outline.

But I wasn't able to make them work perfectly. The black body was out of register around the mouth, front paws, and lower right leg.

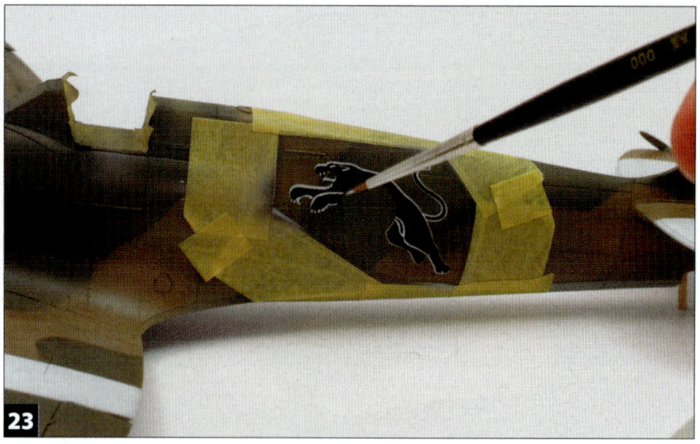

Using Vallejo white and a 000 Series Seven Winsor & Newton brush, I touched up areas around the upper mouth, lower front paws and lower right leg.

The two-color SEAC roundels and fin flashes were applied with a mix of 85 percent light blue (XF-23) and 15 percent) white (for the centers, followed by a mix of 70 percent blue (X-4), 15 percent sea blue (XF-17) and 15 percent medium blue (XF-18) for the outers.

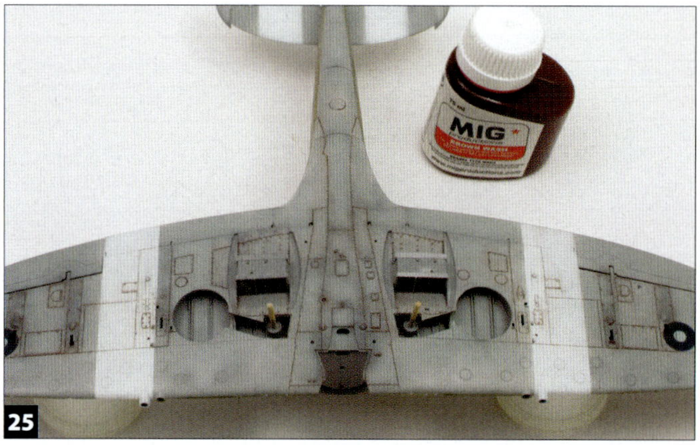

With painting complete, I moved on to weathering beginning with Mig Production's brown wash in areas behind the engine and around the guns.

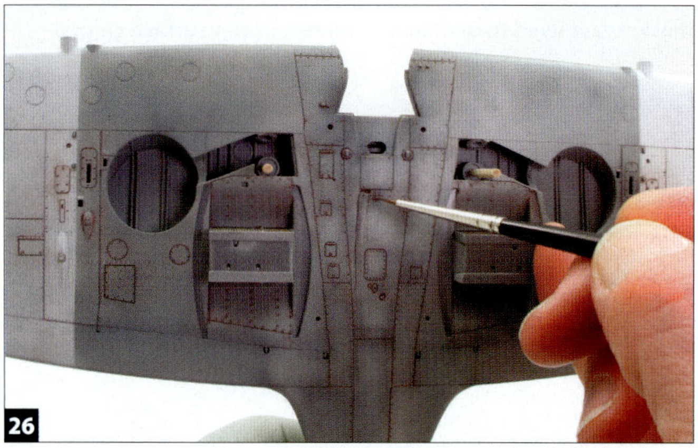

Like most WWII aircraft, Spitfires leaked copious amounts of oil, which streaked back in the airflow. But the oil does not flow straight back. Oil is fairly viscous and it meanders along uneven surfaces, much like water on your windshield when driving in the rain. To capture this look, I mix Winsor & Newton mars black and burnt umber. I begin by applying dabs aft of a given panel line.

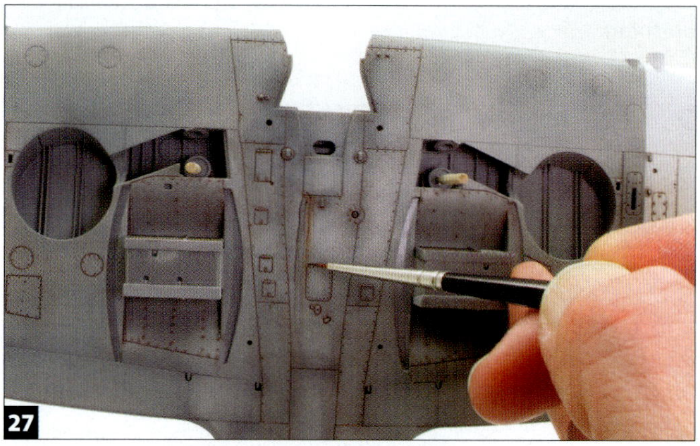

Using a Winsor & Newton Series Seven 000 brush moistened with odorless thinner, I draw the paint rearward. With practice, you'll be able to capture a more natural looking streak.

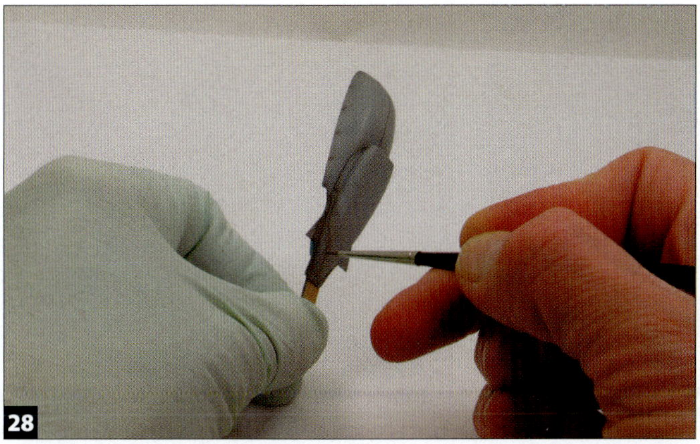
28

Streaks are also added to the lower cowl. Again, I start with a dab of paint at the source of the leak, then with my brush dampened with thinner, draw it back to form the streak.

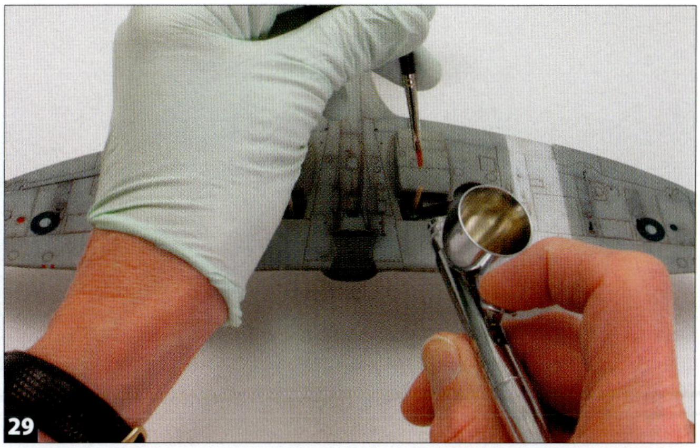
29

I added mud splashes by blowing well thinned 502 Abteilung light mud (ABT 125) with my airbrush from a flat No. 2 brush. Moderation is important as airplanes generally don't get as muddy as tanks, so reduce with thinner any big lumps of paint you may get.

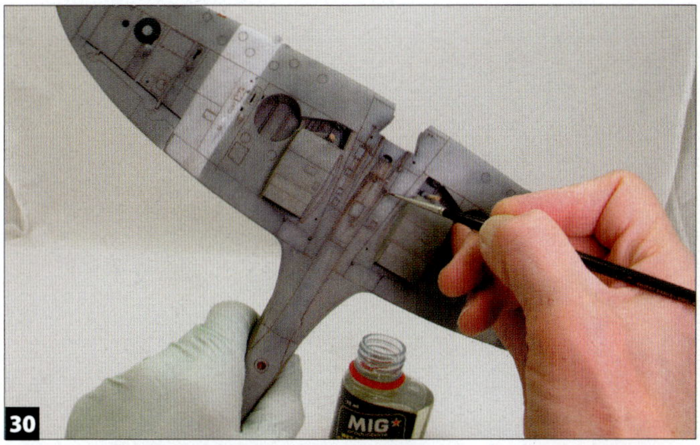

30

As the final step in the underside weathering, I used Mig Wet Effects enamel gloss to represent still-wet oil. A small dab at the start of a streak is applied and then drawn back with a 000 brush.

31

The completed underside should reflect local conditions. In Burma, monsoon rains kept airfields wet, so muddy water would get thrown up from wheels. I used the same 502 Abteilung ABT 125 light mud as I used for the splashes to build up some muddy water marks on the tires. The wheels are BarracudaCast's part No. 32006, drop fit replacements for the unpaintable vinyl kit parts.

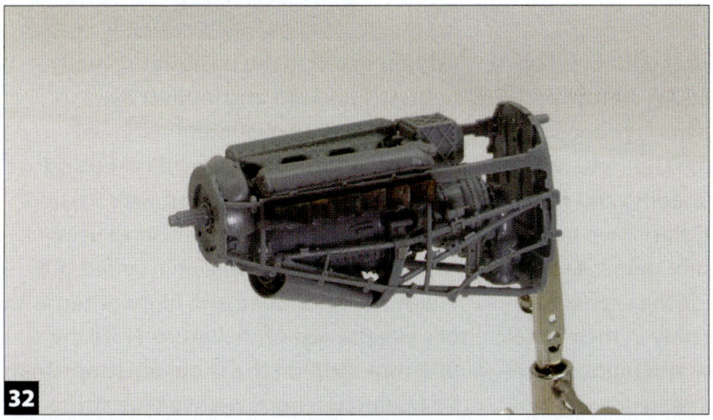
32

I left off all the detail parts on the engine and firewall as I opted not to display the engine. I did incorporate the magnets for the cowls as they will help me position them during final assembly. I replaced the kit exhaust stacks with Quickboost's set QB 32-143, as the kit parts are labor intensive to clean up. They were finished first with Alclad exhaust manifold and jet exhaust and followed by 502 Abteilung oils and Mig pigments.

33

While test fitting for the final airframe assembly, I discovered I had introduced a right-hand twist into the engine assembly and therefore the top cowl was not fitting correctly. To make more room for me to flex the top cowl into a better position, I cut away the top of the engine blocks, parts H6 and H9. I used Zouke-Mura's Pro Zestu Nipper to complete this surgery.

34

The cowl magnets hold these parts for positioning while I used Tamiya's Extra Thin Cement lock the four-part cowl assembly into place.

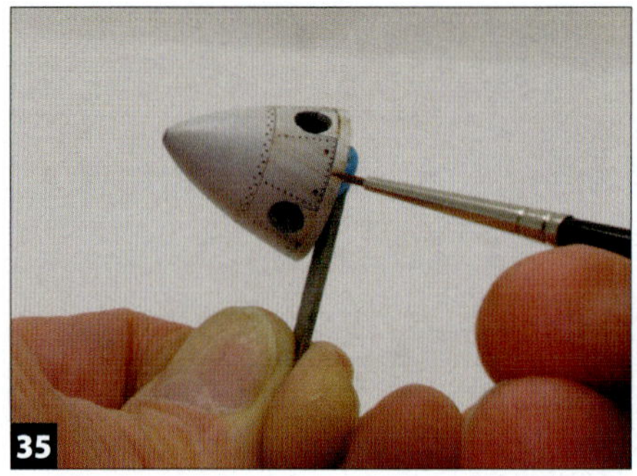

35

Propeller hubs also leak oil, but you don't want to overdo this effect, especially on light colored spinners. Tiny dabs of Winsor & Newton burnt umber are applied at the seam forward of the prop blades, then drawn back with a 000 brush dampened with thinner. Due to centrifugal force, the oil curves gently around the spinner.

36

I replaced the Spitfire cockpit door with BarracudaCast's part No. 32002. Tamiya's rendering is marred by ejector pin marks that are tough to clean up. BarracudaCast gives you a separate crowbar, but they often went missing, so I left mine off. The Sutton Harness is from RB Productions.

37

Brian used the head from Wings Cockpit RAF tropical pilot. The torso, shorts, legs and arms (gray in this photo) are from Dragon's 1/35 scale Commonwealth Infantry set (No. 6380). Brian scraped away the molded webbing detail, then rebuilt the pockets, shoulder straps and the shirt's back and created a pair of gloves using Apoxie Sculpt. The shorts got belt detail made from .050-inch plastic sheet. Shoes from the Wings Cockpit figure were superglued to the legs with short pieces of paper clip inserted into drilled holes. Brian lengthened the legs so the figure was now closer to 1/32 scale. The calves were then resculpted with Apoxie Sculpt to replicate cuffed knee socks. Shoelaces are .15-inch lead fly-tying wire.

38

Brian base-coats his figures with inexpensive Americana and Deco Art craft store acrylics, then uses oil paints (Winsor & Newton, Grumbacher and Old Holland are his favorite brands) for shading and detail work. Here, the face and shirt have been finished with their oil paint details—compare these to the rather lifeless-looking shorts and legs that are only in their acrylic base coat. Initial shadow and highlight colors are first blocked in, their edges blended using a stippling motion with a dry Size 1 brush. Lighter highlights and deeper shadows are then worked in wet-on-wet. Once dry, lightest highlights and deepest shadows are applied (in oils) wet-on-dry with a 00 brush. Each step in the process covers a smaller area with greater contrast.

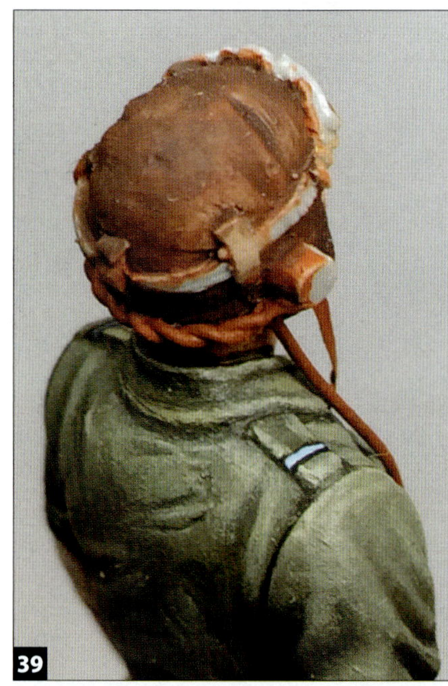

39

Adding detail to the basic figure really helps it stand out and lends it extra authenticity. Brian superglued headphone leads to the flying helmet using two strands of .15-inch lead wire twisted together, and added a chin strap using thin sheet from the seal on a coffee can. These will be shaded and highlighted with oils like the rest of the figure. Note the RAF rank strip on the shoulder strap; Brian first painted a narrow navy blue stripe using acrylics, then used oils and a 00 brush to draw in the light blue inner strip, reducing the dark blue to a narrow border on either side.

40

The oxygen mask and hose came from a spare Ultracast pilot figure, with microphone leads made from lead wire. Minute applications of gap-filing superglue slightly thickened the end of the wire to replicate the red-colored headphone jack. The goggles got several applications of Liquitex clear acrylic gloss to simulate glass lenses. For F/O Ardeline's Errol Flynn-type moustache, Brian used his 00 brush and dark gray oil paint.

41

To help fill large areas of "dead space" around the Spitfire, Brian assembled a chair with the pilot's parachute pack draped over it. He used a resin chair from Plus Models' Easy Line collection (No. EL049), combined with the parachute pack from the Wings Cockpit resin figure. He then sculpted the parachute's back pad and straps from strips of partly-cured Apoxie Sculpt before painting the whole unit in oils over acrylics just like the figure itself.

42

The base is poplar, which is denser than oak and therefore will take a stain better. Brian and I played around with the positioning of the Ardeline figure using my previously built Tamiya Mk. IX as a stand-in for the Mk. VIII Ardeline flew.

43

The groundwork is tile grout with various grades of static grass added along with a few jungle leaves from Ammo. These leaves are paper and therefore much easier to work with than PE as you don't have to worry about paint scratching or flaking off.

44

The groundwork was airbrushed shades of Tamiya browns, red browns, greens and yellow, switching between colors until we were satisfied we had gotten an effect that suggested the monsoon region in which F/O Ardeline was based.

45

To suggest the damp conditions of the far east, the groundwork was finished off with some gloss areas using Pledge Floor Gloss (PFG) and Mig Wet Effects enamel gloss. The PFG is perfect for wet areas, while the Mig Wet Effects is used to suggest dampness. These coatings were only applied to dark earth colored areas.

10 Weathering a strafer

Painting Italeri's B-25G Mitchell for the southwest Pacific

AARON SKINNER

Sometimes, bombers found themselves in a gunship role—and the North American B-25 was a good one, and makes a striking model with its bold nose artwork.

Designed as a medium bomber, North American's B-25 Mitchell proved to be a versatile platform adapted for a variety of roles. One of the most striking applications was the gunship, which replaced the glass nose with a mount for two to eight machine guns and, in the case of the aircraft I built, a 75mm cannon, all firing forward. Many squadrons wrapped the nose with artwork including bats, cats, and dragons.

I built Italeri's 1/48 scale B-25G, an early purpose-built strafer. This kit started life as Accurate Miniatures B-25D released in 1999. Looking over the striking marking options included, I gravitated to one from the 822nd Bombing Squadron

operating in New Guinea in 1943. Of course, as is always the case, reference popped up too late indicating that the subject aircraft was probably fitted with four guns as well as having the 75mm gun replaced by two more .50-caliber machine guns.

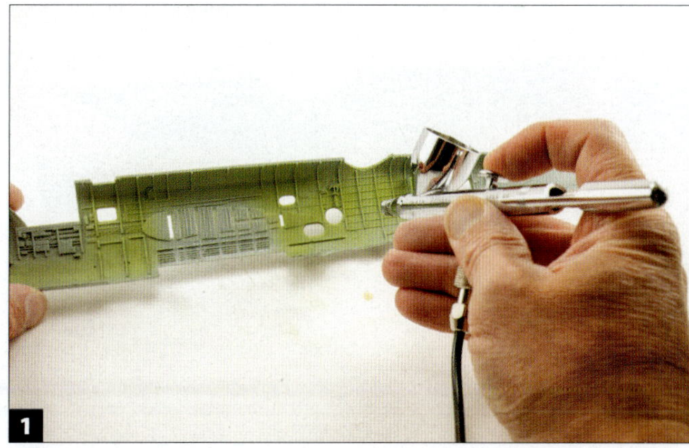

Although much of the Mitchell's interior won't be especially visible through the few windows, I wanted what is there to stand out. I started by spraying much of the fuselage with AK Interactive yellow green (AK2206).

Research indicated the cockpit as probably dull dark green, which I airbrushed with AK's color (AK2106). Stiff paper protected the section behind the framing that will remain yellow green.

To paint the aluminum bomb bay walls, I dry-fitted the bay assembly into the fuselage and held it tight against the side to serve as a mask. I used the same technique to paint the cockpit area.

4

I dry-brushed the raised detail molded inside with a yellow green mixed with white. It's a lighter shade than I would normally use, but the interior will be dark so I wanted to heighten the contrast.

5

Next, I applied brown Tamiya Panel Line Accent Color to deepen engraved detail and recesses. This enamel wash flows easily into recesses, but I brushed it on a little heavier than normal to again dial up the contrast.

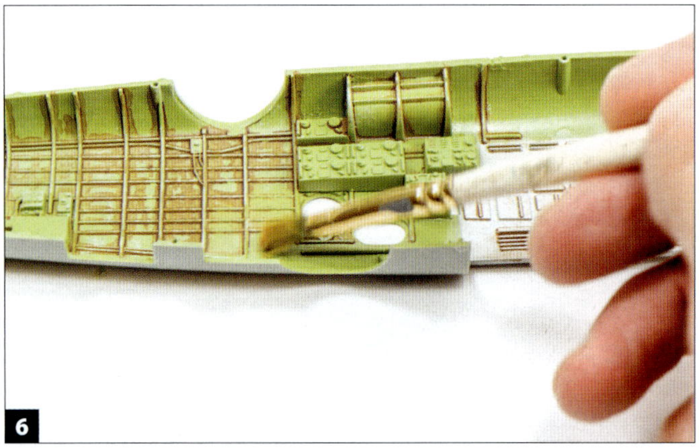

6

To refine the wash, I applied clean enamel thinner with a large soft brush. The thinner reactivated the brown. I blotted excess from the brush between strokes to avoid transferring the darker shade.

7

To mask the turret and windscreen, I pieced together small slivers and triangles of tape to gradually fill each pane.

8

After gluing the windows into the fuselage sides, I carefully burnished tape into the frames with a toothpick, until the outline was easy to see.

9

Then, I traced around the edge with the tip of a new (sharp) No. 11 blade and peeled up the excess. Don't press too hard when tracing the frame with the knife to avoid scratching or marring the clear plastic.

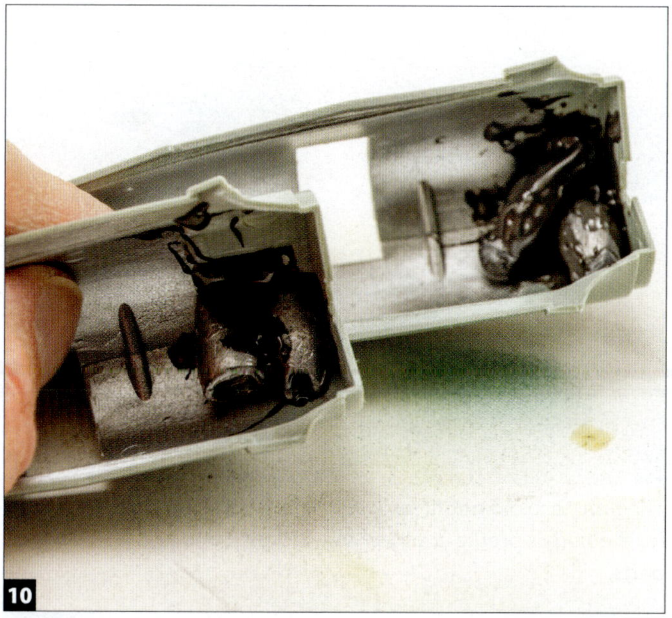

In order to keep the nose wheel on the ground, I filled the front gear bay, lower part of the nose, and the forward part of each engine nacelle with fishing sinkers held in place with JB Weld.

I painted the engine cylinders Tamiya metallic gray (XF-56) and crankcase dark gray (XF-24) with details picked out with flat black (XF-1) and flat aluminum (XF-16). Tamiya black panel line color highlighted detail.

After protecting surrounding detail with strips of tape, I sanded the fuselage seam. Poor fits produced gaps and steps in a couple of places.

Superglue set with accelerator and sanded smooth took care of gaps that sanding alone failed to eliminate.

A fine razor saw guided by tape engraved panel lines lost to sanding. It takes only a light touch and a few strokes to replace this kinds of lost detail.

The wings fit pretty tightly into recesses in the fuselage. I squeezed a little Vallejo Plastic Putty, an acrylic filler, along the seams for a clean finish.

A cotton swab dampened with water smoothed the putty to produce a tight seam.

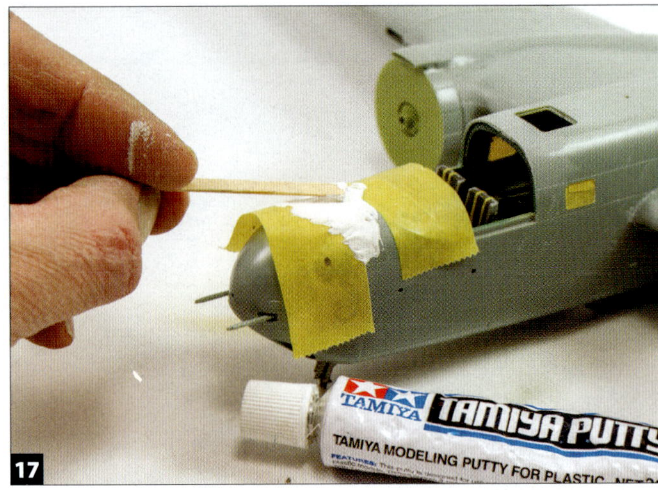

The gun nose did not fit well, so I used Tamiya filler putty to build up the body's profile ahead of the cockpit. Sanding blended the parts.

Before painting, I masked openings for the landing gear, crew ladders, and the astrodome with chunks of foam. Using pieces slightly larger than the openings allowed the foam to expand and plug the voids.

After cleaning the model with rubbing alcohol, I airbrushed the underside of the Mitchell with AK Real Color neutral gray 43 (RC261). Mixing these acrylic lacquers about 50 percent with Real Color thinner and spraying at about 20 psi produced a smooth, even finish.

To begin post-shading, I mixed a little flat black (RC001) into the mix and airbrushed various panels and edges under the airplane.

Then, I added flat white (RC004) to the base neutral gray paint and airbrushed it thinly, initially on panel centers, then in a more random pattern of splotches and dots. I added a little more white, and repeated the splotching to break up the gray.

22

To produce a slightly soft boundary between the gray and upper camouflage, I rolled out worms of poster putty and placed them in a wavy pattern before spraying Real Color dark olive drab 41 (RC259).

23

By spraying across the edge (rather than up under it), I was able to produce a distinct edge with a little softness, as if it had been sprayed freehand at the North American factory.

24

Photos show that the fabric-covered control surfaces look lighter than the metal surfaces around them, either because the paint fades or dries differently. I masked the ailerons, rudders, and elevators, and sprayed them with a mix of olive drab and a little dunkelgelb (RC060).

25

Varying the ratio of olive drab and dunkelgelb, I post-shaded the upper surfaces with splotches and spots of lighter green to mimic sun-fading in the Pacific.

26

Referring to photos, I sprayed olive drab darkened with a drop of black and a little Real Color rotbraun (RC068) along the nacelles on the top of the wings. I believe this is exhaust soiling combined with grime left by ground crew servicing and fueling the plane.

27

Satisfied with the shading, I filled the airbrush with Real Color medium green 42 (RC260) and airbrushed the splotches on the perimeters of the wings and tail planes to match color diagrams. I lightened the color with dunkelgelb to fade the splotches on the fabric surfaces

28

It's not called out in the instructions, but images of 822nd Bombing Squadron Mitchells show yellow on the bottom of the tails. I masked and sprayed flat white as a base before airbrushing Tamiya flat yellow (XF-3).

29

After airbrushing several thin layers of Testors Model Master Metalizer sealer (No. 1409), I applied the kit decals. Four pieces comprised the impressive panther head, which settled over the noses curves with a generous dose of Microscale Micro Sol.

30

I sealed the decals with another layer of Metalizer sealer to lay a foundation for weathering. The first step was to paint Tamiya brown panel accent color along all of the recessed lines and details. I wasn't worried about being neat.

31

After letting the enamel wash dry for an hour, I gently drew a paper towel damp with clean thinner across the surfaces to remove the excess and highlight the panel lines. Thinner on a cotton swab refined the effect in tight spots.

32

To "chip" the paint along the leading edges, I dipped a chunk of soft foam into Tamiya metallic gray, blotted excess off on a paper towel, and then lightly dabbed the foam against the model. The effect should be kept to a minimum and look random.

33

Then I sealed the finish with Testors Model Master acrylic clear flat (No. 4636). Refining the chips, I applied Tamiya flat aluminum with a fine brush both to the leading edges especially around the engines where the prop would throw sand and rocks …

34 … and along panel lines around the engine hatches, gun service doors, and elsewhere. Continuing the notion of the crew walking on the wings, I focused small chips along the grids of rivets molded adjacent to the engines.

35 To soften the chips, blend the camouflage, and add further fading, I misted the upper surfaces with thin Vallejo acrylic matte varnish (70.520) tinted with a drop of AK acrylic dark olive drab 41 (AK2201).

36 After painting the propellers flat back with yellow tips, I gently dry-brushed flat aluminum across the leading edges to add a little wear and chipping.

37 Once the landing gear was on and I would need to handle the B-25 less, I brushed AK black pigment (AK039) back from the exhaust ports on the nacelles.

38 I used a Molotow chrome marker to shine the solid plastic lights on the wingtips and vertical tails as well as the three identification lights molded under the starboard wing.

39 Then, I painted the wingtip lenses with Tamiya clear red (X-27) and clear green (X-25); the ID lights were painted with clear red, clear green, and clear yellow (X-24).

11 Hairspray on a Peggy

Hasagawa's 1/72 scale Japanese bomber gets an old-school chipping treatment

CHUCK DAVIS

Most of Japan's World War II bombers look dated and stodgy in comparison to their Allied counterparts. An exception was the Mitsubishi Ki-67 "Peggy," probably Japan's best medium bomber of the war. Sleek lines and a reasonable armament weren't enough to compensate for low production numbers, however. Unusually—and likely a nod to its capability—while the Peggy was a Japanese Army design, it was also seconded to the navy for use as a torpedo bomber.

I had always admired the looks of the Peggy, and Hasegawa obliged a few years ago with a beautiful 1/72 kit, so I decided to tackle a torpedo-armed version. I thought it would be the perfect canvas for an attempt at using the "hairspray" technique to apply the chipping so associated with Japanese warplanes.

Looking over the kit's contents, I decided to add an Eduard Color photo-etched metal (PE) interior set. Also available—thankfully—is an Eduard mask set, quickly becoming one of my favorite aftermarket add-ons. Surprisingly, there aren't really any other details sets out there dedicated to the Peggy, but I ran across detailed resin weapons for the Hasegawa Emily flying boat kit. In an effort to up the detail further, I decided to give these guns a shot.

The Pacific Theater was a harsh environment, so weathering plays a large role in making models set in the area believable. There are many techniques to get the desired effects, and Chuck turned to a classic for this build.

1

Hasegawa's Ki-67 Peggy is a fine kit, but a few extras can lift the detail and make painting easier. I've added resin guns and color PE parts from Eduard, brass cannon barrels from Master, and an Eduard masking set.

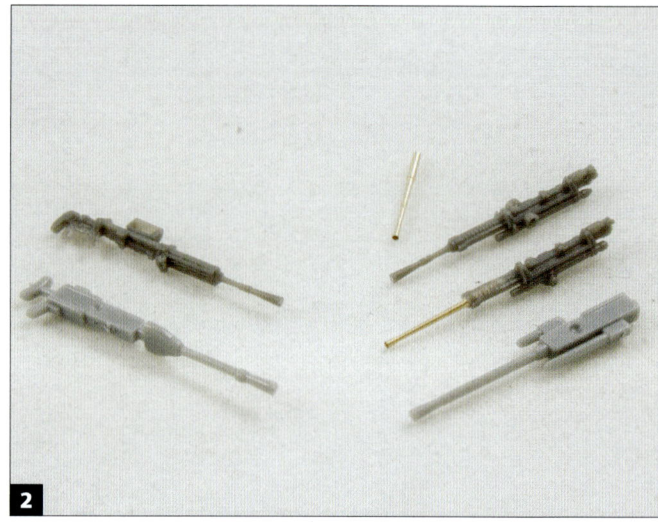

2

One of the more visible ways to add detail is to upgrade the guns. While I couldn't find a set dedicated to the Peggy, I took a shot at using guns from an Eduard set designed for the Hasegawa Emily kit. Surprisingly—or not—the guns and mounts all fit the Peggy very closely. The less detailed kit weapons are obvious.

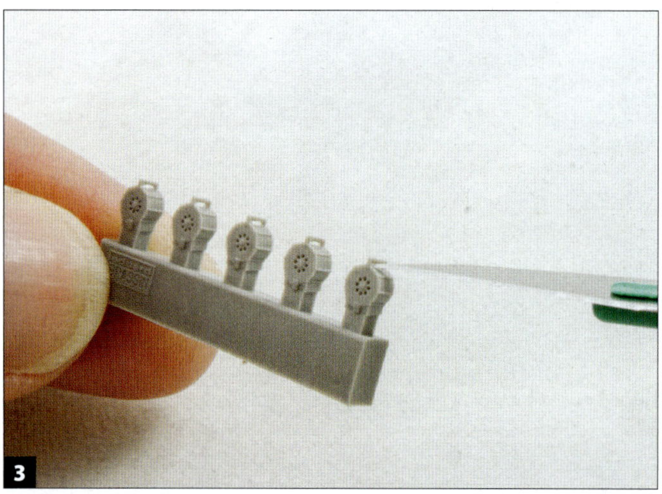

3

The Eduard set includes resin magazines. I enhanced the already nice detail by carefully opening up the handles.

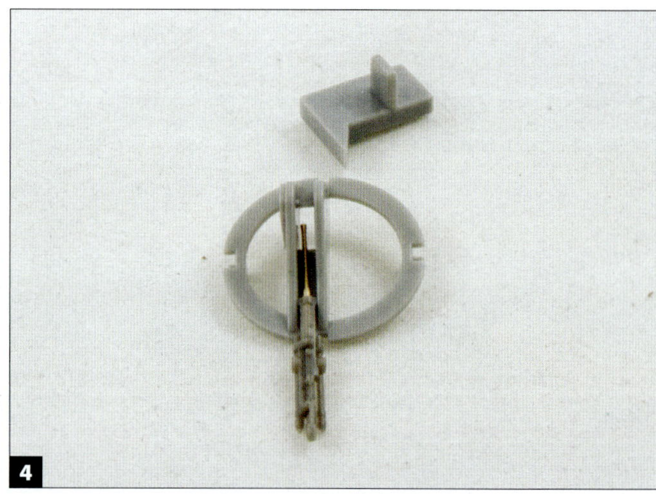

4

A dorsal turret gun mount that was slightly too narrow was the only problem I encountered fitting these guns.

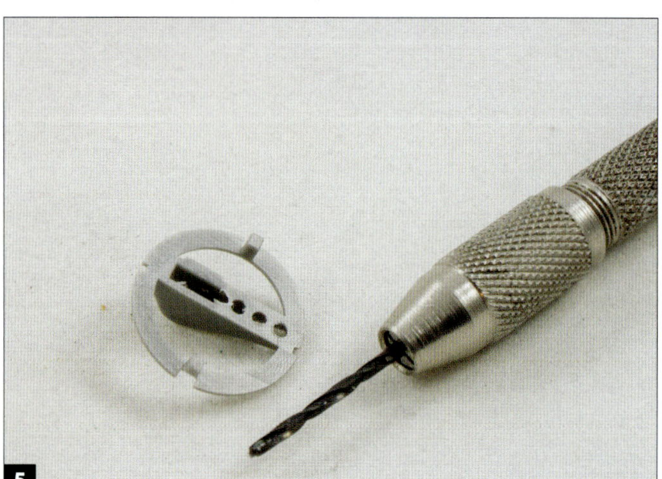

5

To adapt the mount, I "chain drilled" holes in the lower tray connecting the gun mounts.

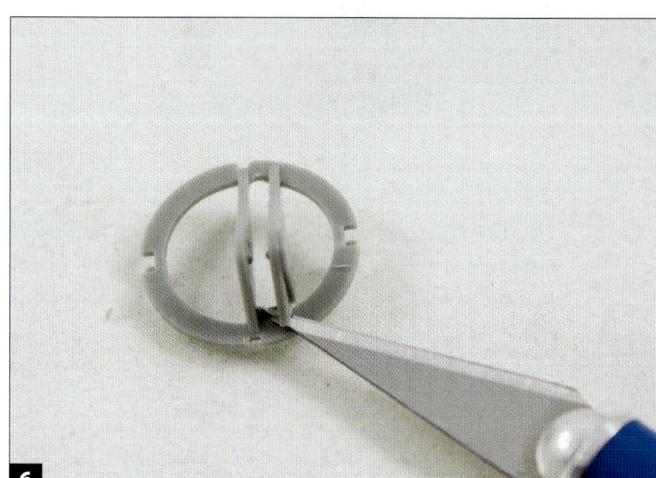

6

This allowed me to cut away the tray and carve the gun mounts off the turret ring. A sharp blade and care are required.

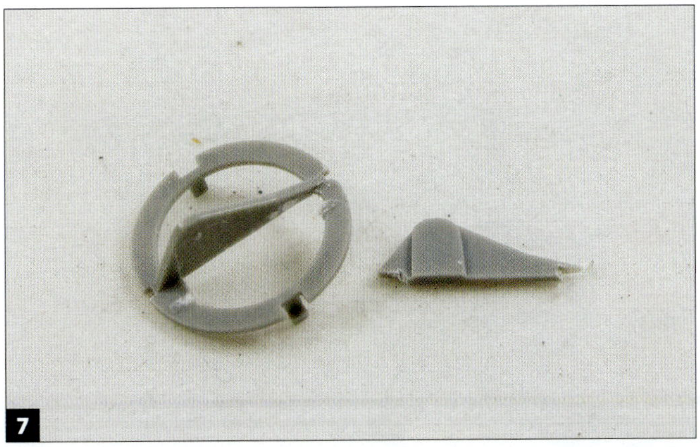

7 I left the center mount in place and removed the outer mount, reasoning that this left enough room for the gunner.

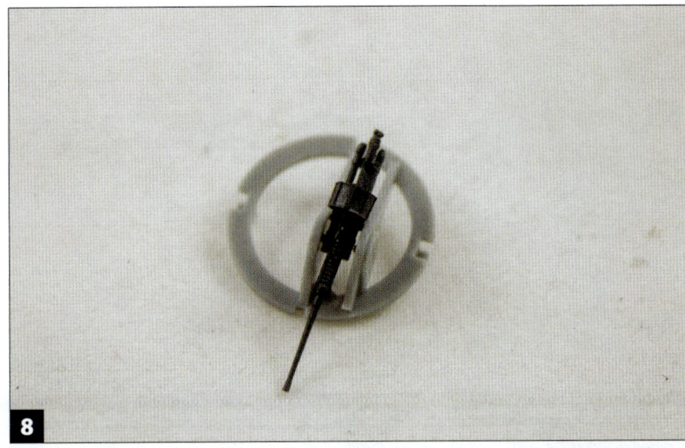

8 Checking the fit showed the skoosh more room worked. You'll need to check the glass to make sure the slot for the gun won't need trimming as well. Mine didn't, but I had to re-mount the gun lower in the cradle so the magazine would clear the top of the turret.

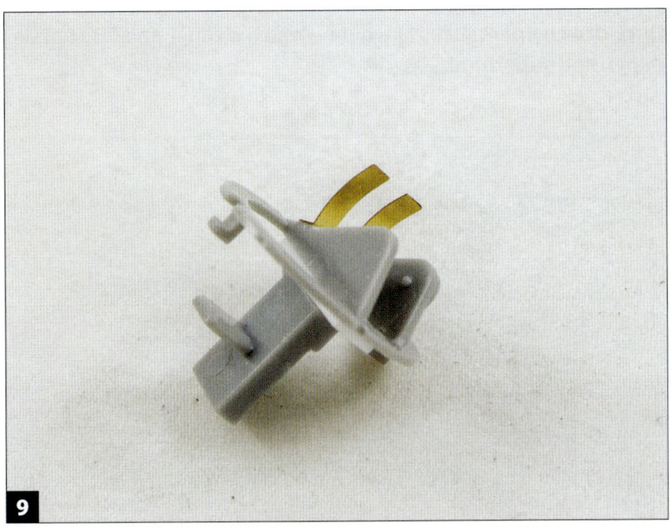

9 I replaced the lower tray with a piece of strip styrene and mounted the PE rear armor plate included in the Eduard set. This results in a reasonably detailed turret—good enough inside a 1/72 scale bubble.

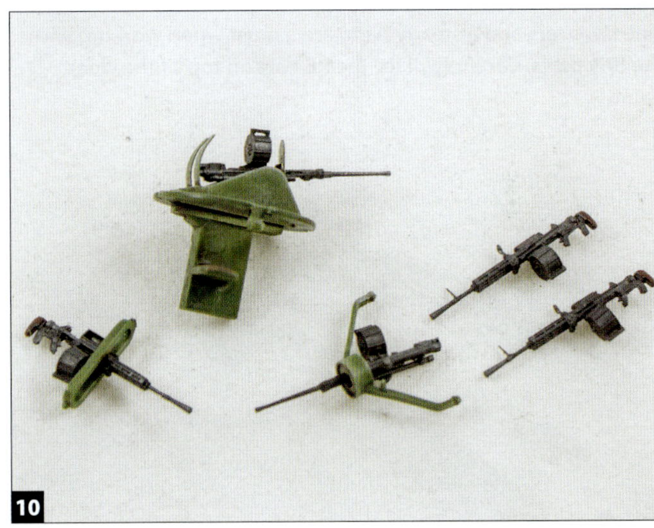

10 After painting and weathering, the enhanced detail of the new weapons is evident.

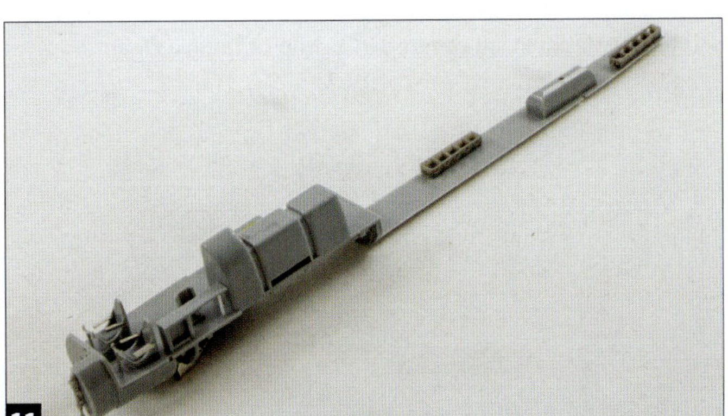

11 Although I didn't have any specific references for the rear fuselage interior, I added the ammo can racks from the Emily set in what seemed logical locations for the rear and waist positions. The Eduard interior set also includes tiny arm rests for the seats. I added them prior to painting as they are quite fragile.

12 Using clear paint is a handy method for assembling tiny PE pieces. It's especially useful where there is some risk of getting glue inside instrument faces—the clear can actually double as simulated glass.

13 Fine tweezers and a steady hand are a must when working with the tiny parts. Carefully place the PE part on top of the clear paint.

14 Once dry, the part sticks like glue—much cleaner and less hassle than using superglue.

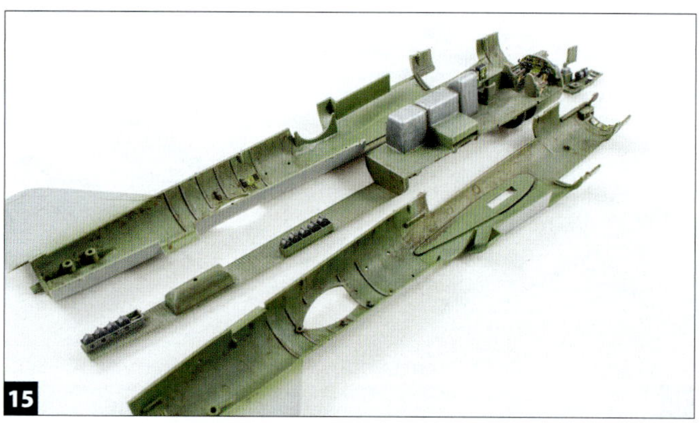

15 The addition of the Eduard PE adds enough detail to be satisfying without being overwhelming or complicated.

16 Some stippling and weathering with Flory Models dark dirt wash adds some interest. I also thinned the bombardier's stool in the nose since it was pretty thick and cushy looking. There is a round protrusion on top of a "box" in the lower nose framework—I made a simulated compass by painting white over black and scratching detail in, all covered in a drop of Tamiya clear.

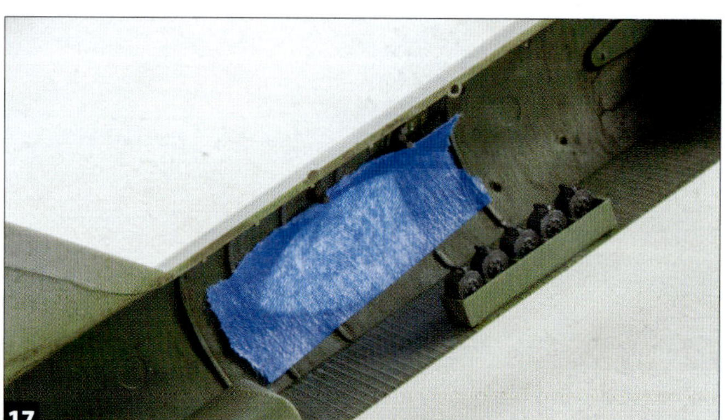

17 Planning ahead, I placed pieces of tape over the open waist positions to make masking easier. They will be easy to remove after all painting is complete.

18 Here's a bit of a throwback—the prop tips have a line molded in place for the tip decals. Those need to go—I'm pretty sure they weren't that way in real life.

19

The Eduard masking set includes pieces for the wheel hubs, but I prefer to paint the wheel and then spray the center hub using a circle template as a mask.

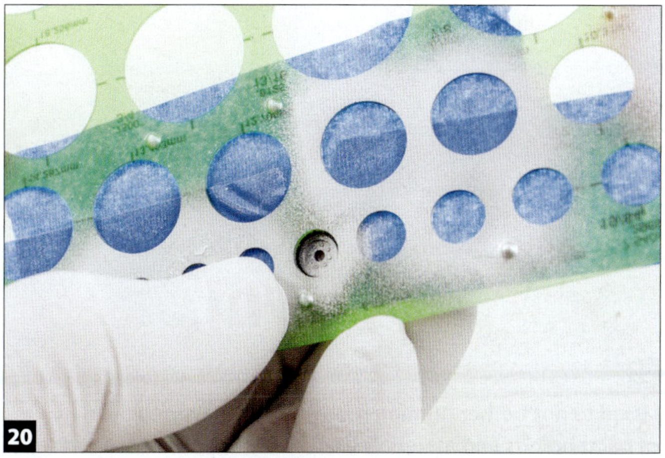

20

Simply find a circle with an equivalent diameter to use as a mask and cover the adjacent ones, then spray the hub color while holding the wheel tight to the back of the template. Hint—pick one side of the template to spray through and always use that side to avoid rubbing paint off on future projects.

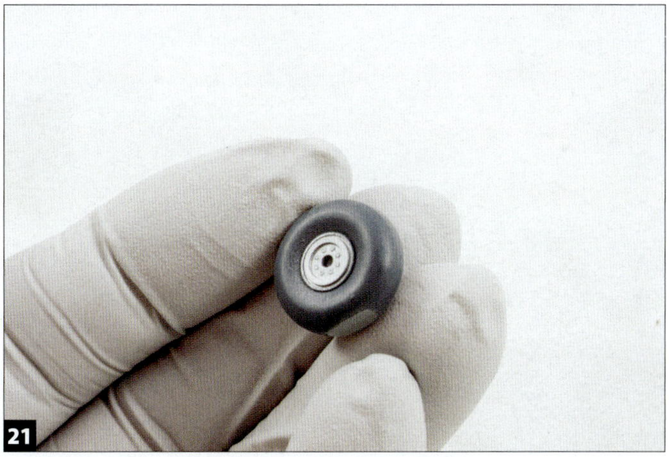

21

Quick and easy—a perfect wheel hub ready for a wash of Flory dark dirt.

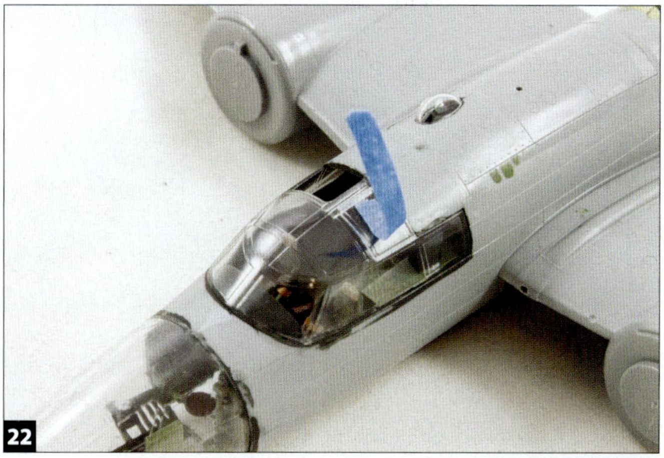

22

Another old-school technique I use is painting the edges of the clear parts with black paint to cut down on unrealistic reflections. You can see what I mean on the edges of the upper canopy panel that I forgot to paint. The piece of tape is there to hold the part during assembly, so it doesn't get dropped into oblivion inside the fuselage.

23

The reflections disappear after the edges are addressed.

24

After applying the canopy masks—bless you, Eduard!—I sprayed the interior color over the exposed frames so it's visible when looking inside. Don't forget to block off the slot for the turret mounted cannon. I used a small piece of sponge.

25

I used to not bother priming my models, but with the variety of paint brands and materials used in kits now I've changed my habits. My current favorite is Alclad gray Microfiller Primer.

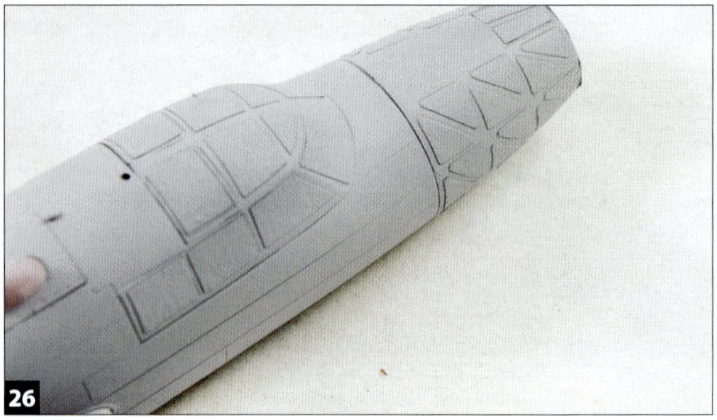

26

And here's the proof that priming can help—even though I thought the clear nose glazing fit well, I was surprised to see a large step appear under the paint.

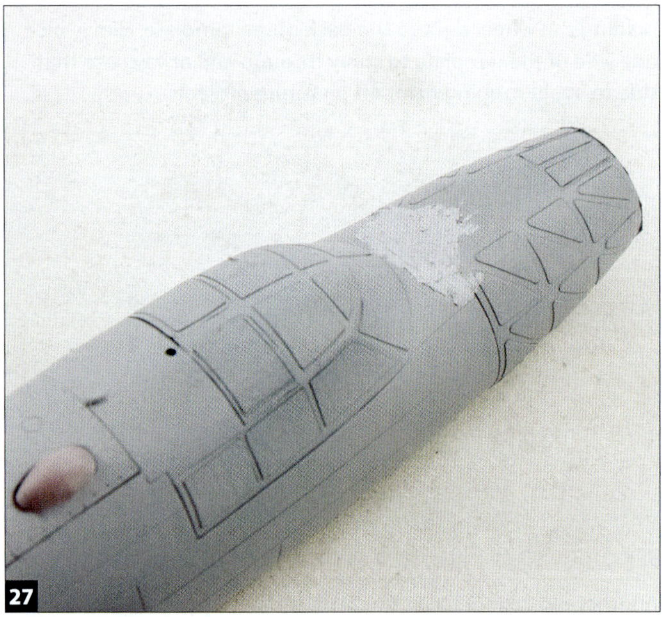

27

Careful application of putty was needed to avoid damaging the canopy masks.

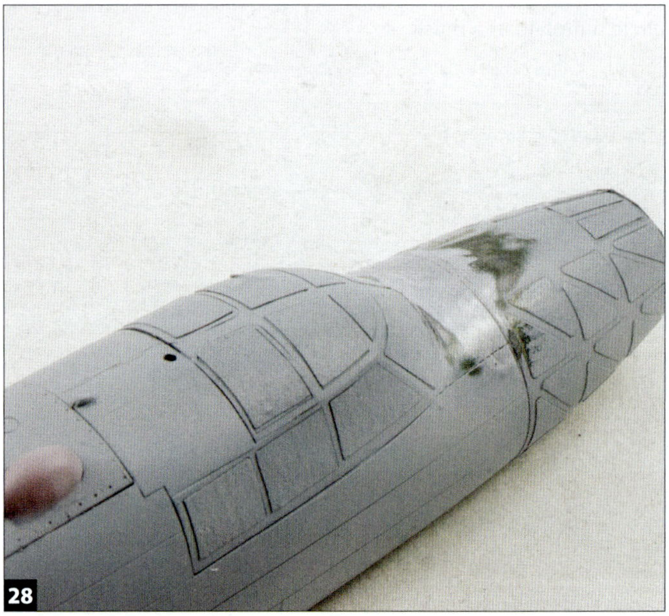

28

Even more care was required while sanding the putty back. Thankfully, the masks stood up well to the beating. A quick re-scribe of the panel line and we were back in business.

29

I masked and sprayed the national markings since they are simple circles. This allows them to be easily shaded and faded just like the rest of the paint. A patchy coat of Tamiya flat red was used directly over the gray primer—the gray dulled the red to a nice faded tone. I also sprayed the white band on the tail.

30

I cut masks using a sharp blade mounted in a drafting compass. The wing markings needed two strips of tape because of their width. The small rectangular pieces in the center of each circle are there to cover the tiny hole made by the compass tip—you can never be too careful about overspray.

31

After masking the markings, I primed again with the Alclad gray to bring everything back to a neutral starting point for the final finishing.

32

Japanese aircraft from WWII are notorious for showing a lot of wear and tear. I wanted to experiment with adding chipping using the hairspray technique. This puts a barrier of hairspray between colors which, when reactivated with water, disturbs the paint adhesion enough to allow scraping and chipping. The first step is applying the color you want to appear beneath the scrapes—in this case, Testors Metalizer steel.

33

The lower surfaces were sprayed with Tamiya J.A. gray (XF-14). After masking that, a patchy coat of Tamiya J.A. green (XF-13) mixed about 70/30 with flat brown (XF-10) was applied.

34

Next, a mix of more J.A. green and a lot of flat white (XF-2) was used to give a random faded appearance.

35

I borrowed my daughter's hairspray (shhhh, don't say anything) and spritzed some on where I had applied the steel earlier. After letting the hairspray dry about 20 minutes, I misted some water on the desired areas using the airbrush and then attacked it with a stiff nylon bristled brush. Just as advertised, the paint scraped and chipped away randomly.

36

I could have extended the trial to the rest of the airframe, but The Peggy doesn't seem to have been worn as much as other types, so I limited the chipping to the engine and center section areas.

37 I removed the masking of the lower surface and national markings and sealed everything in with a coat of Testors Acryl clear gloss.

38 Next up were the few decals needed followed by another coat of gloss—in retrospect, I should have painted the yellow orange leading edges to avoid using the kit decals. Next time.

39 Back to old school again—I applied a wash made with Winsor & Newton burnt umber oil paint thinned with Turpenoid.

40 After waiting 30 minutes or so, I used a paper towel to scrub the oil paint off in the direction of the airflow.

41 The gloss coat not only protects the paint but also keeps the chipped areas from chipping more.

42 When possible, I like to spray the gear wells and other interior openings after applying the overall camouflage colors since its usually easier to mask around the edges of the wells than mask inside them.

First, a base coat of Testors steel is applied to the gear wells and the bomb bay.

Next, a thin overcoat of Testors interior metallic blue adds the correct tint without turning the wells a shocking blue.

With the painting finished, it was time to add the final pieces to this war-worn Peggy.

12 *Tico* ready for launch

Camouflaging Trumpeter's small-scale Essex*-class carrier*

AARON SKINNER

1 Trumpeter molded the hull above the waterline as a single part with internal braces, so it's solid. But there are mold seams at the bow and stern to be removed.

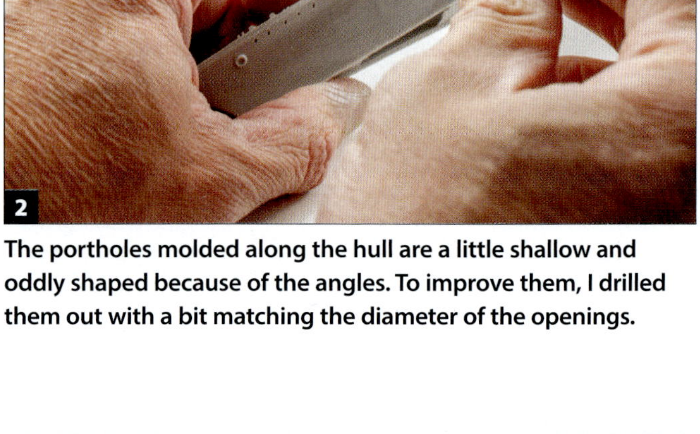

2 The portholes molded along the hull are a little shallow and oddly shaped because of the angles. To improve them, I drilled them out with a bit matching the diameter of the openings.

If one weapon sets the Pacific War apart, it is arguably the aircraft carrier. From the moment Japan used them to strike Pearl Harbor in 1941, aircraft carriers would be at the center of naval battles and support amphibious operations.

In the drive to defeat Japan, the U.S. Navy built 24 *Essex*-class carriers that could embark 90-100 aircraft, travel 20,000 nautical miles, and were protected by dozens of antiaircraft guns.

USS *Ticonderoga* (CV-14), was 16 feet longer than its predecessors to accommodate bow guns and entered service in mid-1944. What attracted me to Trumpeter's 1/700 scale kit of the ship was the attractive camouflage shown on the box. When the ship entered service, it wore Measure 33 Design 10A camouflage, a series of geometric shapes on the hull and superstructure.

Dazzle camouflage and a varied airwing make Essex-class carriers such as the USS Ticonderoga *interesting modeling subjects.*

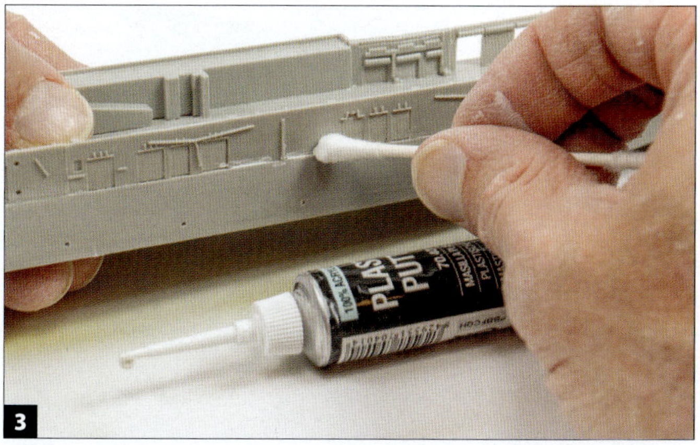

3

I filled gaps where the hangar-deck inserts met the hull with Vallejo water-based putty. Smoothing it with a damp cotton swab minimized cleanup.

4

While most of the kit's ejector-pin marks will be hidden on the finished model, there are a few that need to be taken care of, especially under deck overhangs and a few of the larger gun tubs.

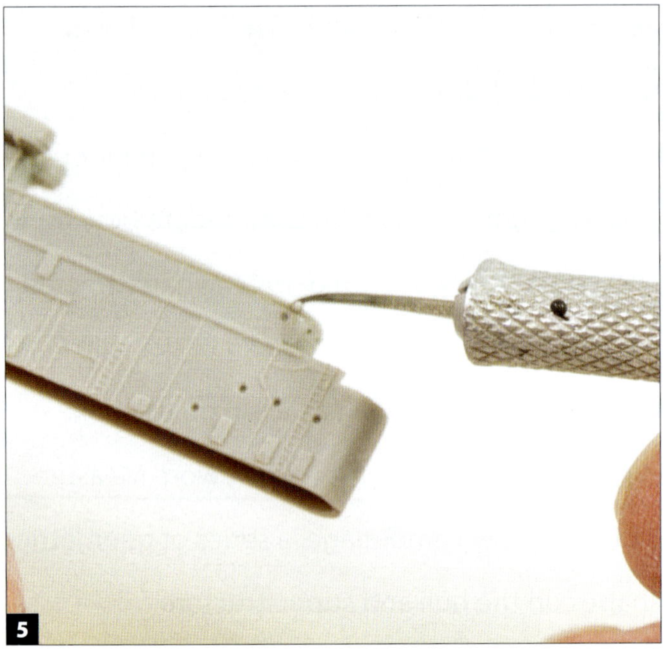

5 The fit of some of the island parts was less than perfect, so I filled gaps with superglue flowed in from a Glue Looper.

6 Looking to upgrade the detail a little, I picked up Tom's Modelworks *Essex*-carrier photo-etched metal (PE) details (No. 757). As well as railings and radar array, the set included gun shields.

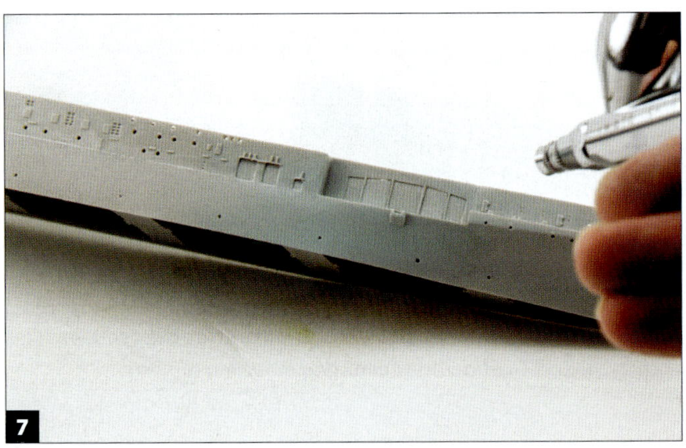

7 Once the hull was together and cleaned up, I airbrushed it with Testors Model Master 5-L light gray (No. 4863) enamel starting in the recesses and then applying it in clean, smooth passes.

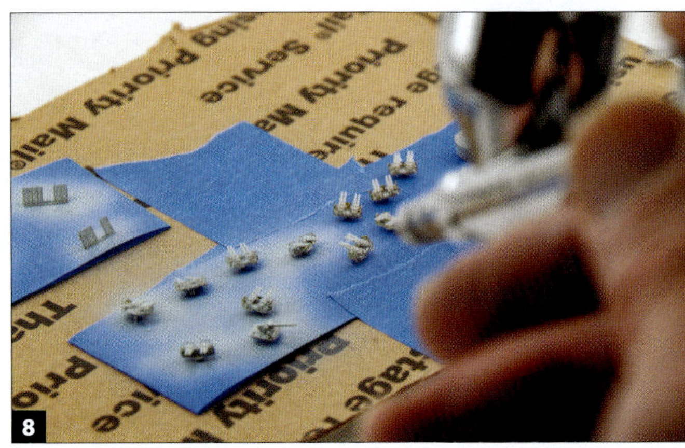

8 While I had light gray in the brush's reservoir, I painted previously assembled guns and masts as well as small individual parts still on the trees ...

9 ... and the Tom's Modelworks' PE frets I taped to cardboard for support. Remember to flip the fret later to ensure both sides get painted.

10 After letting the light gray dry for two full days, I began masking. I enlarged plans for the Measure 33 Design 10A camouflage scheme to match the model and better replicate the pattern.

11

To match the shapes on the starboard side, I used Tamiya tape for curves. Like electrical tape, this flexible vinyl stretches just right to form smooth curves.

12

Then, I sprayed the middle color of the camouflage, 5-0 ocean gray, using enamel from True North paints (No. 2011). Keeping the mix thin and spraying at a low pressure prevents paint building up along the tape edges.

13

Two days later, I masked the ocean gray with strips of Tamiya tape.

14

Next, I applied the final and darkest of the camouflage colors, 5-N navy blue, another True North enamel (No. 2026).

15

The next day, I carefully pulled up the layers of tape to reveal the geometric disruptive pattern.

16

After hand-painting the decks and gun trough floors with True North 20-B deck blue (No. 2013), I scraped paint away from attachment surfaces and glued on additional gun decks and tubs.

17

With the hull painted, I moved on to the deck, first airbrushing overhangs and vertical surfaces 5-L light gray, then airbrushing the deck with 20-B deck blue.

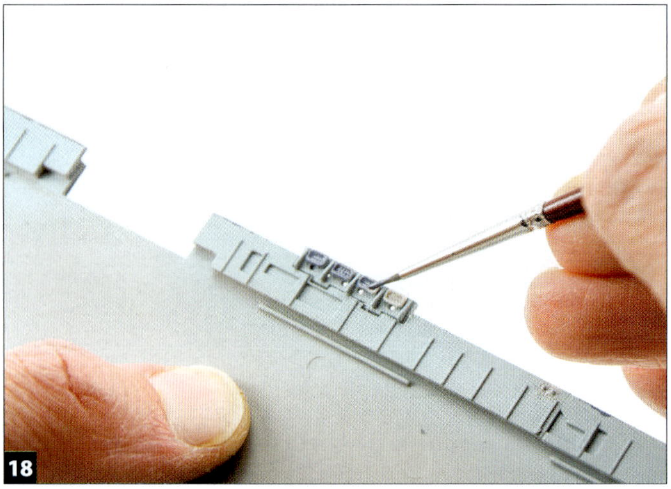

18

I glued the Carley floats under the gun tubs and painted them 5-0 ocean gray.

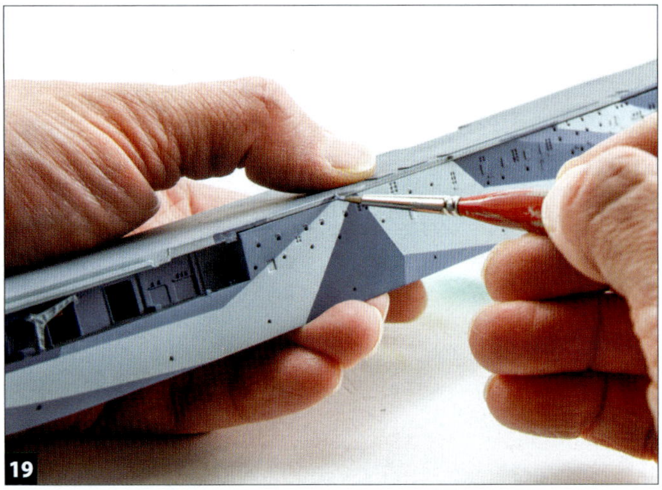

19

Rubber bands and clamps ensured the deck fit tight to the hull as the glue set. Then, I painted the camouflage up the sides of the deck piece to match the diagrams.

20

After spraying the model with Testors Metalizer sealer, which produces a thin, clear gloss, I applied the kit decals. Most of the sheet provides stripes and elevator surrounds for the fight deck.

21

I wanted to secure the hull to the vacuum-formed seascape base that comes with the kit before adding delicate parts. First, I airbrushed the underside of the wake and bow waves seafoam green mixed from Tamiya acrylics.

22

Then, I airbrushed a wider area around the wake and waves with an intermediate blue shade, still from underneath.

23

The final color applied from under the base was Tamiya sea blue. When flipped, the flat paint on the back side of the textured plastic looks like glossy water.

24

The enhance the wake, I dry-brushed the peaks with Tamiya white to give the effect of saltwater foam churned up by the ship.

25

I wanted to be sure the ship stayed on the flexible base, so I attached it with Dap Alex Plus clear caulk. It sets better than white glue and any excess pushed out of the recess will look like water.

26

After pressing the hull firmly into the base — brace it from underneath when you do this or the ship won't make full contact—I placed a couple of books on the deck and left the caulk to set for a couple of days.

27

With the ship in place, I could handle it by the base. That meant delicate parts being added, such as the scores of antiaircraft guns around the deck, were in less danger from my clumsy hands.

28

Although the guns had been painted before installation, many still required touch-ups with a fine brush and 5-L light gray.

29

Moving topside, I masked and airbrushed first 5-O ocean gray, then 5-N navy blue on the island. The same colors were hand-painted on smaller item like gun turrets and range finders.

30

I also hand-painted 20-B deck blue onto upper surfaces and decks. I thinned the paint and let it flow around the edges of gun tubs, but it then required multiple coats for proper coverage.

31

As I added decks to the ship's island, I hand-painted the railings to match the camouflage pattern.

32

Before adding fragile subassemblies, such as radar dishes and antennas, to the island, I glued it to the deck.

33

The final step for the ship was adding fine PE railing and the radio masts at the edge of the flight deck with thin superglue.

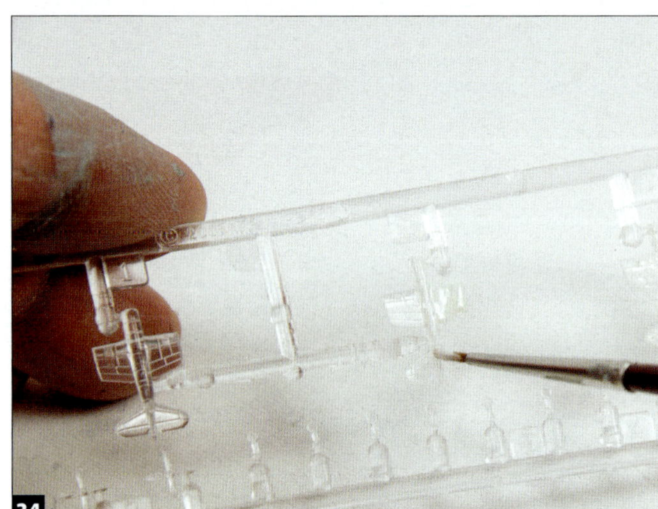

34

A carrier means aircraft and the kit provides six TBF Avengers, nine SB2C Helldivers, and 15 F6F Hellcats. Photos indicate the airwings were camouflaged differently. I painted the bombers in three-tone white, intermediate blue, and dark sea blue.

35

But the fighters were finished in overall gloss sea blue. All aircraft were painted by hand on the trees using Testors Model Master enamels. Note that I cut most of the wings so they could be posed folded on deck.

36

To properly mark the airwing, I used Starfighter Decals USN generic air group markings sheet (700-104). Magnification was essential to get these tiny insignia aligned.

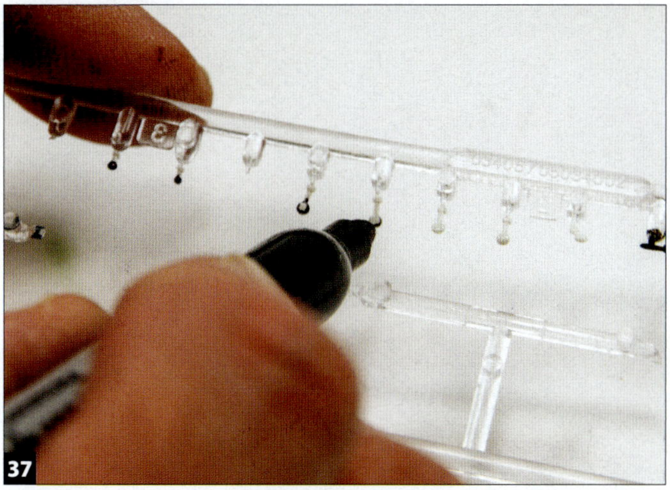

37

Separate landing-gear legs and propellers are provided lending scale finesse but also upping the ante on the complexity of assembly. I used a Sharpie to draw tires on the landing gear.

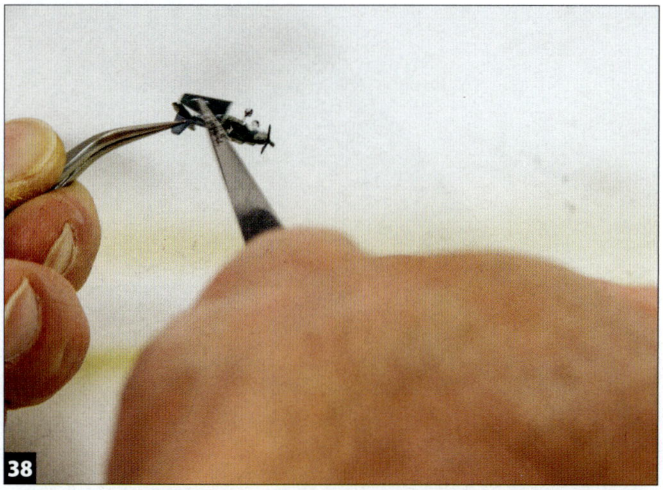

38

I carefully reattached the wings in the folded position using dots of superglue, two pairs of tweezers, and plenty of muttered curses. A couple of aircraft were lost to legs and wings that squirted from the tweezers, never to return.

39

And several Hellcats had to be left off after I inadvertently applied insignia to the wrong side. Let's just say that several of the fighters have already launched on a mission.

40

Using dabs of superglue on each of the wheels, and tweezers to position the aircraft, I stored most of the aircraft on deck. Three Hellcats with wing extended are preparing to launch.

With its air wing complete, the USS *Ticonderoga* steams into the wind for another mission.

Building an LVT-4 for Okinawa

13

AFV Club's 1/35 scale LVT-4 just needs judicious finishing to make a standout model

ANDY COOPER

The Landing Vehicle Tracked (LVT) was a series of military tracked amphibious landing vehicles spawned from an American private venture to develop a rescue amphibian for use in the swamps of the Florida Everglades. The vehicle was powered in the water and on land by its uniquely designed tracks and was originally known as an Alligator in its civilian guise.

The U.S. military soon realized its potential as a landing vehicle and development quickly produced a line of military variants, the most numerous being the LVT-2 and the improved LVT-4. These types of vehicles became known as Amtracs, as they are still called today.

For my model I portrayed an LVT-4 that took part in operations on Okinawa in April 1945 with the Army's 7th Infantry Division. The LVT-4, with the engine moved to the middle of the vehicle and a rear loading ramp fitted, allowed improved access and egress to and from the cargo hold and also allowed the LVT-4 to carry a small vehicle such as a jeep.

This is a more or less an out-of-the-box build of the AFV Club kit (No. AF35205) and the color scheme of this particular LVT-4 caught Andy's eye as being different from the usual olive drab or gray overall schemes.

1

I kicked off by making the wheel assemblies. With 20 units required, each consisting of four parts each, this was a tedious task, but there were no issues here. Just be aware that one wheel on each side fits straight into the hull part and does not need to be assembled as a complete bogie. The drive and idler sprockets were also completed at this stage and put aside for later.

2

The supplied interior parts for the driver compartment are detailed and well molded, with a transmission, steering levers, seats, instrument panel, fuse boxes, air filters, and more all included in the kit. Hardly any of this can be seen when the compartment is closed up, but in order to populate the inside, it is worth doing, especially if you intend to pose the hatches open.

3

To make fitting and painting the driver compartment easier, I glued the floor to the lower hull. A little weight helped keep the load compartment floor and rear wall in the correct location while the glue set.

4

The internal side walls of the driver compartment need to be painted before final assembly. These are the forward portions of the hull (parts C11 and C12). I masked off the relevant areas, sprayed them with Tamiya gray primer, and later airbrushed the parts flat white.

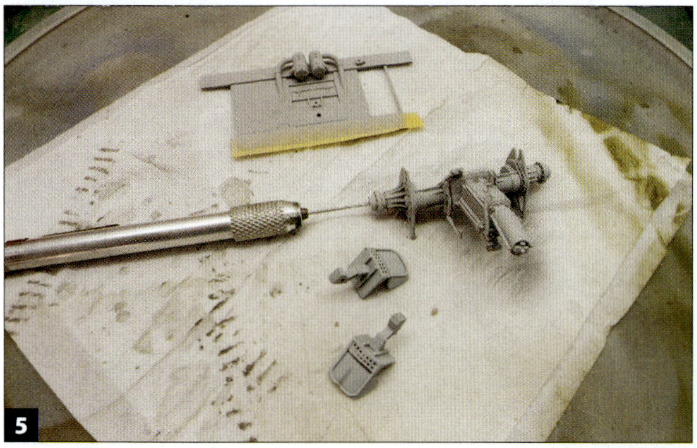

5

I prepped the individual interior components with Tamiya primer using an airbrush.

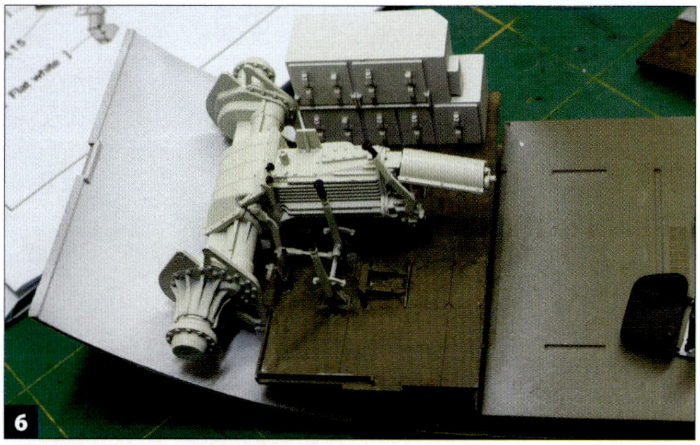

6

I sprayed the interior parts white. To add variation, I used Hobby Color off white (H21) on some components such as the transmission. Once dry, the major subassemblies were attached. The compartment floor (B43) is not white, but the same color as the remainder of the vehicle. I used lightened olive drab.

7

I prepared and assembled the hull sides next as these are an integral part of the completed driver compartment, and I wanted to ensure they were as complete as possible before putting everything together. This process included attaching the previously made boogie assemblies, and some judicious clamping until the glue set.

8

Dry-fitting the hull sides using clamps showed everything was going to fit, but at this stage only one side was glued into place. This is due to more complexity of the ongoing painting process of the engine compartment behind the driver. The engine sits in the middle of the vehicle with a passage on each side of its compartment to gain access to the cab and for engine maintenance. The walls of these passages need finishing and painting before the major components of the hull can be glued in place.

9

I removed the unglued side of the hull, and masked the engine space passage. I sprayed a base coat of olive drab (Hobby Color H304). I used a lighter color than the instructions specified because upcoming oil washes would darken the finish to near the shade specified.

10

At this point I started to weather and wear the interior compartment using dark gray and a small brush to create paint scratches. Paint wear was enhanced further with the use of a small piece of plastic pot scrubber dipped in the same gray paint and dabbed in the high traffic areas using tweezers to hold the pad. I gave the remainder of the interior components, including the rear wall (part C6) and the compartment roof (parts F12, C1), the same treatment.

11

I completed the outer walls of the engine space passages and painted them. The large red fire extinguisher in the right hand passage adds a splash of color.

12

I sealed the paint and weathering in the driver compartment with Alclad Aqua Gloss prior to a light application of a black oil wash. With that complete, the time had come to attach the second side to the hull and begin to close up the driver compartment.

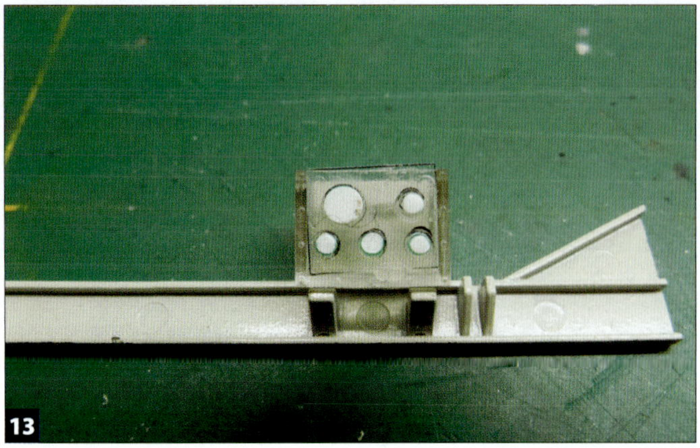

13

Masking the clear molded instrument panel and its film dials posed an issue. In order to spray it white, I punched out correctly sized discs of thin styrene sheet, then secured these over the dial faces with a drop of masking fluid.

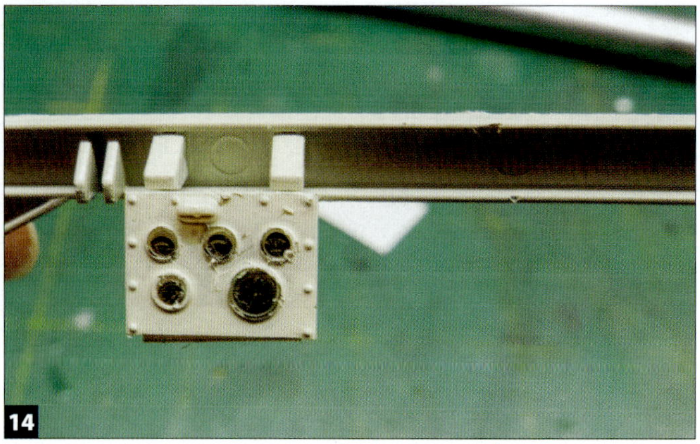

14

After spraying, I removed the discs, revealing the instrument faces. After a clean up, I glued the instrument panel assembly into the driver compartment.

15

The remainder of the engine space passages were assembled before adding the load space wall (Part F13) in preparation for the roof, which covers both these spaces and the driver compartment.

16

The driver compartment roof consists of several parts. I found it easier to assemble the flanks and front plate (Parts F9, F10 and F6 respectively) to part F12 with slow-setting glue before adding the hatch opening panel (Part C1). I assembled this as described and sprayed the inside surfaces flat white. Rubber bands and a clamp held everything in place for the final assembly as the glue set.

17

Small gaps were evident, and I filled these with a smear of Tamiya putty before sanding it to the profile of the hull.

18

Sections 30-32 of the instructions deal with the characteristic "squashed honeycomb" trim vanes on the rear of the vehicle. What is a little confusing in the instructions is that there are two parts with the same number required, numbered A10 for the left vane and A11 for the right vane assembly, not one as depicted and numbered. These parts are left- and right-handed, so be sure to get things the correct way around.

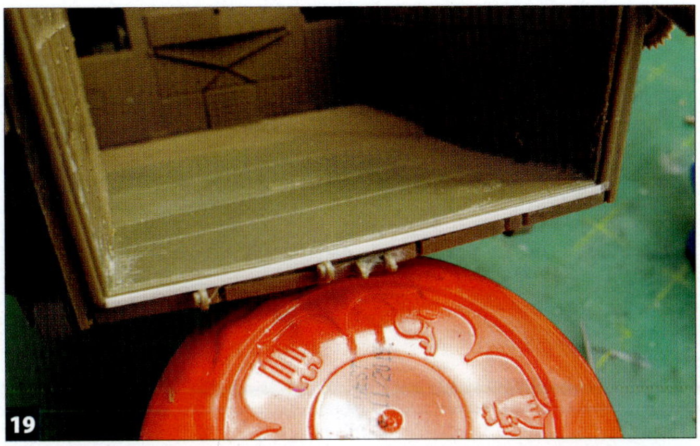

19

I posed my ramp closed. However, there was a considerable gap at the hinge point. Assuming not even U.S. Marines would contemplate boarding an amphibian that was sure to sink like a stone, I closed the gap with a strip of styrene.

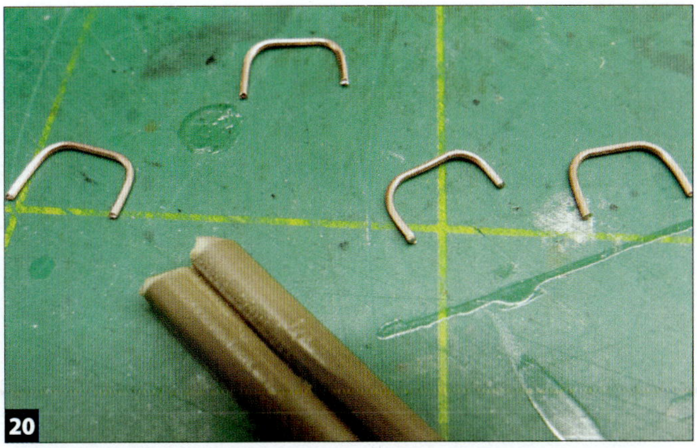

20

The LVT has many grab handles adorning its structure. All of these are supplied in styrene from the box, but I always knock them off and lose them, so I now make such things from thin copper wire, in this case strands of data network cable stripped of its insulation. I used the kit parts as a size guide and glued two pieces of suitably sized sprue together, and made a form sized for most of the handles.

21

I left the locating ends a little longer to be superglued into appropriate holes drilled on the model. I fabricated smaller hooks and other similar fittings from wire and glued them in place. The end result was that all those otherwise small delicate parts were replaced with robust alternatives that survived the remainder of the modeling process.

22

I coated the clear domes over the periscopes and the headlight lenses earlier with acrylic furniture polish and left them to harden. I masked the domes with thin strips of Tamiya tape and masking fluid before gluing them to the hatches before painting.

23

The model was now ready for primer. I used Tamiya gray primer from a bottle thinned with lacquer thinner and applied with an airbrush. Pieces of packing foam in the hatch openings protected the interior from overspray.

24

Once the primer had dried, I applied the base coat of olive drab, as mentioned earlier (Hobby Color H304). With the base coat on and dry, I masked the interior with a "wall" of painters tape and used Scale Model Supply (SMS) masking putty to lay out the camouflage pattern for the first color, in this case Hobby Color sandy yellow (H79).

25

The putty masking was repeated for the second camouflage color of brown. I used Tamiya NATO brown (XF-68). This is the first time I used the SMS putty, and it sticks to just about everything it touches, and when warm will droop off the model. That said, it was useful and easily removed without staining or damaging the paint work. Overspray did not remain on the putty either, but I do not really know where it went!

26

With the camouflage finished, I sealed the model with Alclad Aqua Gloss. Then I applied the four decals required. I resealed with Aqua Gloss once the decals set.

27

I fitted the tracks at this point so they would be included in the weathering process. However, they broke in a couple places as they are quite delicate even though they are vinyl tracks. As the tracks stick with regular model cement, I used scrap styrene strips to rejoin the breaks. Black paint and some thought applied to where the repair would sit made the repairs almost invisible.

28

To start the wear process, I added chips and scratches using a small brush and a dark gray acrylic paint. Some metallic gray was also used to simulate heavier wear, such as on the front of the vehicle that often appears scraped in period photos.

29

I added chips and scrapes on the sides of the model using dark gray acrylic paints and a small brush.

30

The next part of the process is adding run off and staining. Application of artists oils in small dots were streaked downward using mineral spirits to create the subtle look. I used burnt sienna, raw umber and titanium white in varying degrees for this process.

31

Next came some mud build-up and splatters. Suitable water color paints mixed with fine sawdust to give it some body were applied randomly to the running gear and the sides of the vehicle.

32

The final stage was to tie everything together with an overspray of thin acrylic earth brown to give the impression of grime and dust buildup.

33

The interior of the cargo hold came in for some similar treatment. Though I posed my ramp closed, it should be remembered that when open it would receive considerable punishment as men and machines were unloaded. With painting and weathering complete, I sealed the model with clear flat acrylic to take the gloss off and seal the finish.

34

The last thing to do to complete the model was to add the machine guns and their associated shields and glue the hatches in place. I posed the hatches open to reveal some of the interior. However, the white interior color on the open hatches struck me as too stark, so I hand-painted them olive drab, reasoning the crew would possibly have painted the inside surface to make an open hatch less obvious.

The LVT in all its forms has become synonymous with the campaigns of World War II in the Pacific, in particular the amphibious operations to re-take the islands from the Japanese. There is no doubt this line of vehicles played an important role in achieving those objectives by being able to cross coral reefs that would block conventional landing craft from reaching the beach.

Sands of Iwo Jima

14

A photo of a Sherman tank disabled in the black sands inspired this vignette

JOHN BONANNI

The battle of Iwo Jima is arguably one of the most epic battles of the Pacific Theater. Stories from the island transformed America and the flag raising on Mount Suribachi is still a celebrated image to this day. When researching the battle I found a photograph of a Sherman bogged down in the iconic black sand formed by volcanic ash. The Sherman, named *Tokyo Express*, was from the 5th Tank Battalion and became disabled when shrapnel punctured its radiators. The more I studied the photograph the more details popped, making it perfect for a vignette.

The small scene is centered on Dragon's M4A3 Sherman, which was heavily modified to incorporate all of the unique features added by the Marines.

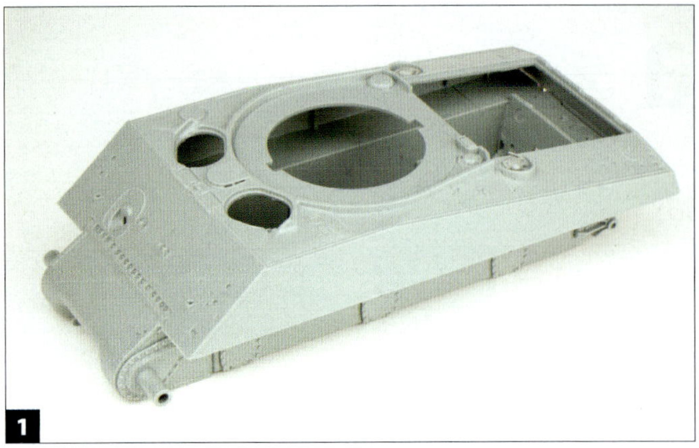

1

Dragon's hull is very straightforward and alignment was perfect. I left the engine deck panel off because I was switching the kit's M4A3 Sherman configuration for an M4A2.

2

I assembled the suspension quickly, and I wasn't terribly concerned with seams because a lot of the details will be hidden once the vehicle is sunk into the sand.

3

A later style commander's hatch was sourced from the spares box since one was not included in the Dragon kit. I filled and textured the turret seam using Tamiya basic putty thinned with Tamiya extra thin liquid cement, and applied with an old brush.

4

The appliqué wood planks for the sides of the vehicle were a combination of balsa wood strips and coffee stirrers, attached with 5 minute epoxy.

5

A tricky feature to replicate on this vehicle was the unique corrugated sheet metal with a jagged edge. To accomplish this I made a simple press with alternating styrene rod on the top and bottom, then sandwiched a piece of lead foil between them. I cut the sharp edges using a hobby knife.

6

I used Aves Apoxie Scuplt, a two-part, water-based putty, to make each sand bag. Once the basic shape of each bag was formed, it was placed on the model and refined with an old paintbrush, a rubber-tipped brush, and a sculpting tool.

7

The single-piece resin rear wading trunk was made by Value Gear Details. I added a photo-etched metal (PE) screen on top for crisper detail and framed it with styrene strips.

8

The tedious and long task of adding the crew improvised anti-personnel spikes was done with stretched sprue. I tried several methods for this feature, and stretched sprue proved to be the easiest. Liquid cement provided ample working time and the end result was resilient to handling.

9

I added additional Value Gear stowage items, along with spare track links from Panda. I sculpted the mantlet cover using Aves Apoxie Sculpt. Now the Sherman was ready for primer.

10

I used One Shot primer from Ammo by Mig Jimenez on the tank. The water-based acrylic polyurethane went on smooth and revealed any errors made during construction.

11

One detail I missed were the bolts used to attach the wood planks. I fixed this easily with bolts from my spares box I had saved from old kits. I took this opportunity to add damage to the wood as well.

12

Vallejo Model Air Russian green was the first layer of green added to the model. At first I liked this color but as you'll come to read, it was resprayed with a lighter, more appropriate, olive drab.

13

I used masking putty from Ammo, similar to Silly Putty, to mask large patches of the vehicle. This product can be reused countless times and doesn't leave any residue, sometimes found with Silly Putty.

14

Throughout my research, I found several color illustrations of this vehicle and one thing was consistent, they were all different. Each illustrator had a different viewpoint on the vehicle's colors, which included red, yellow, tan, and green. I opted for a three-tone scheme. The first color added was Ammo's U.S. modern vehicle tan. It appears very bright, but it will be toned down during the painting and weathering process.

15

As mentioned earlier, the first green added to the vehicle just didn't look right. To appease my reservations I airbrushed Ammo's olive drab base over the darker green patches. The result was exactly what I was looking for.

16

The last camouflage color added was medium rust from Ammo. This color is mostly artistic license, some vehicles had shades of orange on Iwo Jima but as mentioned earlier there's a lot of speculation for this vehicle.

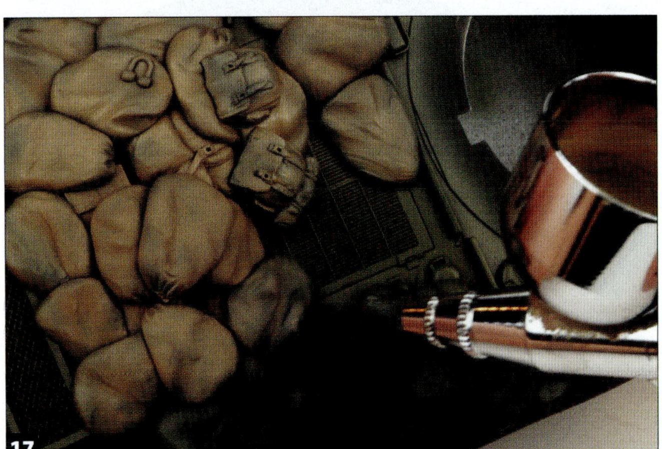

17

Next, I blocked out stowage items with an airbrush. This was easier than using a brush and helped create subtle tone variations that laid the basis for shadows.

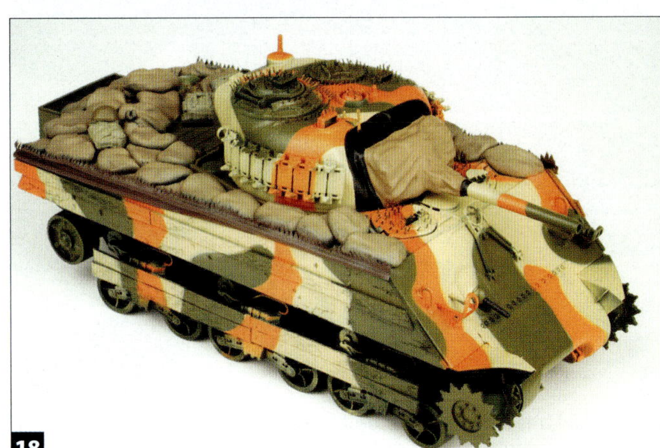

18

I touched up any overspray with a brush and added the distinctive tar sealant around the mantlet cover and barrel with flat black.

19
The model is ready for weathering. At this stage everything has been painted and now the fun can begin. No varnish was used thus far during the painting process.

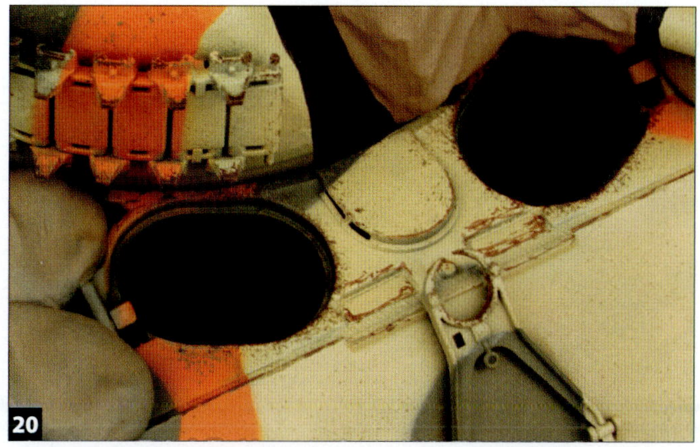

20
I began chipping with a dark brown color applied with a sponge. I returned with a fine-tipped brush to touch up any areas that didn't look in scale. An added benefit of using acrylic paint during this process is that if a mistake is made it can be easily removed with a hobby knife blade.

21
Two birds with one stone. After applying a pinwash around small details, I cleaned up the wash using an overall filter. This helped accelerate the build because typically I wait 24 to 48 hours after applying a filter to add a pinwash.

22
I first added shadows to the sand bags with dark acrylic brown tones. At this point I gave the paint chips some dimension with a lighter olive drab color, which was focused around the dark brown.

23
Sand bag highlights were from the ochre tone range, which kept the overall warm hue. I also picked out details on stowage items.

24
Color tones similar to what was used on the sand bags helped add shadows and highlights to the mantlet cover. In this case I worked rather quickly using three highlight and three shadow color combinations.

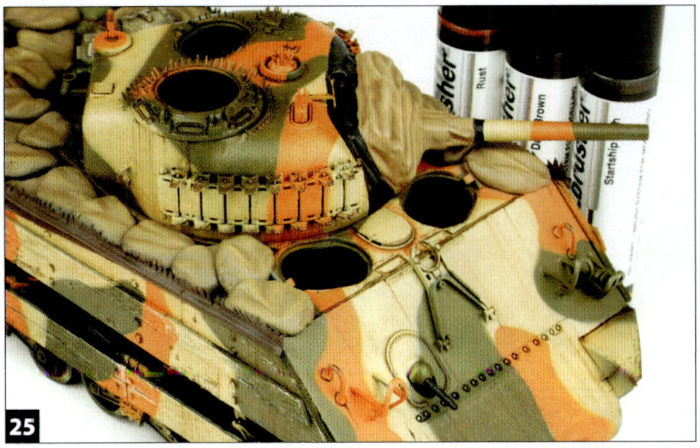

25 I added darker streaks and stains with three colors (rust, dark brown, and Starship Filth) from the Ammo Oilbrusher range. This is a relatively new product and performs very similar to oil paint and can be thinned with mineral spirits, enamel thinner, or Turpenoid.

26 I created the first general layer of dust by airbrushing the vehicle's entire upper surface with AK Interactive's Kursk earth enamel.

27 As soon as I finished airbrushing, I flooded the surface with Turpenoid and began to wear away the dust color. I used a stumping motion to create random patterns, and streaking helped draw the dust down vertical surfaces.

28 A 1:1 mix of Ammo gunmetal and aluminum provided the track base coat—don't worry, the brightness will all but disappear after weathering.

29 I love telling small stories with models and the lid of an unfinished jewel box served as the perfect base. The side walls were balsa sheets and the center was filled with extruded-foam insulation board

30 Prior to finishing the ground work I stained the base, followed by several coats of fast-drying polyurethane. Once dry, the finish is rock solid and the glossy surface allows easy cleanup if some paint or modeling compound get on it.

31 After a building out the center with foam, I used Sculptamold modeling compound for the generic ground contours. The compound leaves ample working time and dries solid within a few hours.

32

Once the vehicle was mounted to the base, I added another layer of Scupltamold, giving the vehicle a sunken appearance.

33

I then spread Vallejo black lava acrylic paste over the white Scupltamold, and before it dried, graded sand was sprinkled on top to enhance the texture. Immediately afterward I turned the base upside down to shake off excess sand.

34

Next I took the equipment scattered around the base and depressed it into the sand for a natural look. Once the acrylic paste dries, these items can be popped off because the sand acts as a barrier.

35

After about six hours of drying time, I masked the wood sides with painters tape and the entire base received a coat of flat black.

36

Iwo Jima sand is black in appearance, but when examined more closely it has a brown cast to it. To mimic this I airbrushed light coats of flat brown from oblique angles.

37

After examining the photo of *Tokyo Express*, I noticed the vehicle had a lot of tarps in addition to just sand bags. I returned with Aves Apoxy Scuplt, which I rolled into thin sheets with the aid of baby powder. Next, the large sheet was cut to size and folded on the vehicle, and once dry I popped them off and painted them.

38 Painting was straightforward with varying green and brown base coats being airbrushed on. A few tarps were treated with the USMC camouflage pattern.

39 I added shadows to the tarps with Oilbrusher Starship Filth. This grimy gray color was perfect and worked well with all of the stowage base colors.

40 The unique trees in the background of the photo were not commercially available. I modified Woodland Scenics trees by removing branches and texturing them with Tamiya basic putty.

41 I added a subtle bark texture with Rustoleum Mutlicolor Textured spray paint. Additionally, I cut individual leaves from laser-cut palm tree fronds, and placed them on the end of each branch.

42 Building up the dense island vegetation began with a layer of 12mm and 7mm static grass applied with the Static King, a static grass applicator made by Woodland Scenics. A mix of water and wood glue mixed to a milky consistency bonded the static grass to the base.

43 I built up multiple layers of laser-cut Hart's Tongue plants from Ammo and bent them to shape. A single drop of super glue bonded all of the layers together.

44 I threw on random tufts, dried sea grass, and more static grass to create the dense vegetation found on Pacific islands. I also added several bushes from Martin Welberg Scenic Studios, which are simply fantastic.

45

Small details scattered around the base really make the scene; in this case ammo cans, PE ammunition belts, and a gun tripod were sourced from the spares box.

46

A final step is to pick out the periscopes by first painting in flat black followed by a top coat of Ammo Oil Stains. I've found this enamel paint and has much better capillary reaction properties than acrylic paints. I simple touched the center of the periscope and the gloss enamel flowed across the entire area.

47

I added a personal element with the addition of a beautiful multimedia figure from Andrea Miniatures. Minimal cleanup was required on the white metal and resin pieces.

48

Painting was accomplished with acrylics. The first step was to prime in black, then I shot white from above, and lastly I added a light uniform coat of olive green. This simple technique highlights all of the folds and provides a solid color foundation. I reproduced flesh tones by mixing these three Vallejo colors: arena marron, basic skin tone, and black red.

49

Despite dry-fitting the figure, he wouldn't stand straight on the base. An easy fix was to add more earth texture to make him appear to be standing in the sand as opposed to just resting on top.

15

Fade to black

Techniques to keep an all-black P-61 night fighter from looking like a toy

CHUCK DAVIS

One of the most innovative aircraft to see service in World War II, the Northrop P-61 Black Widow is a favorite. Although many Widows saw service in standard olive drab (OD) over gray, I think the aircraft looks best in black. The challenge with modeling in an overall one-color scheme is to avoid making it look toy-like—this is especially true for an all-black aircraft.

I selected Great Wall Hobby's 1/48 scale P-61B Black Widow, along with a host of aftermarket items to liven up the detail and add some corrections. While my Widow is basically black, there were lots of things in my favor for avoiding the "toy" look—even my selected scheme of *Cooper's Snooper* from a KitsWorld decal sheet set the (skin) tone. The large amount of glass in the P-61 allows the interior to play a big role in avoiding the toy look, and the Eduard photo-etched metal (PE) detail set adds plenty of eye candy.

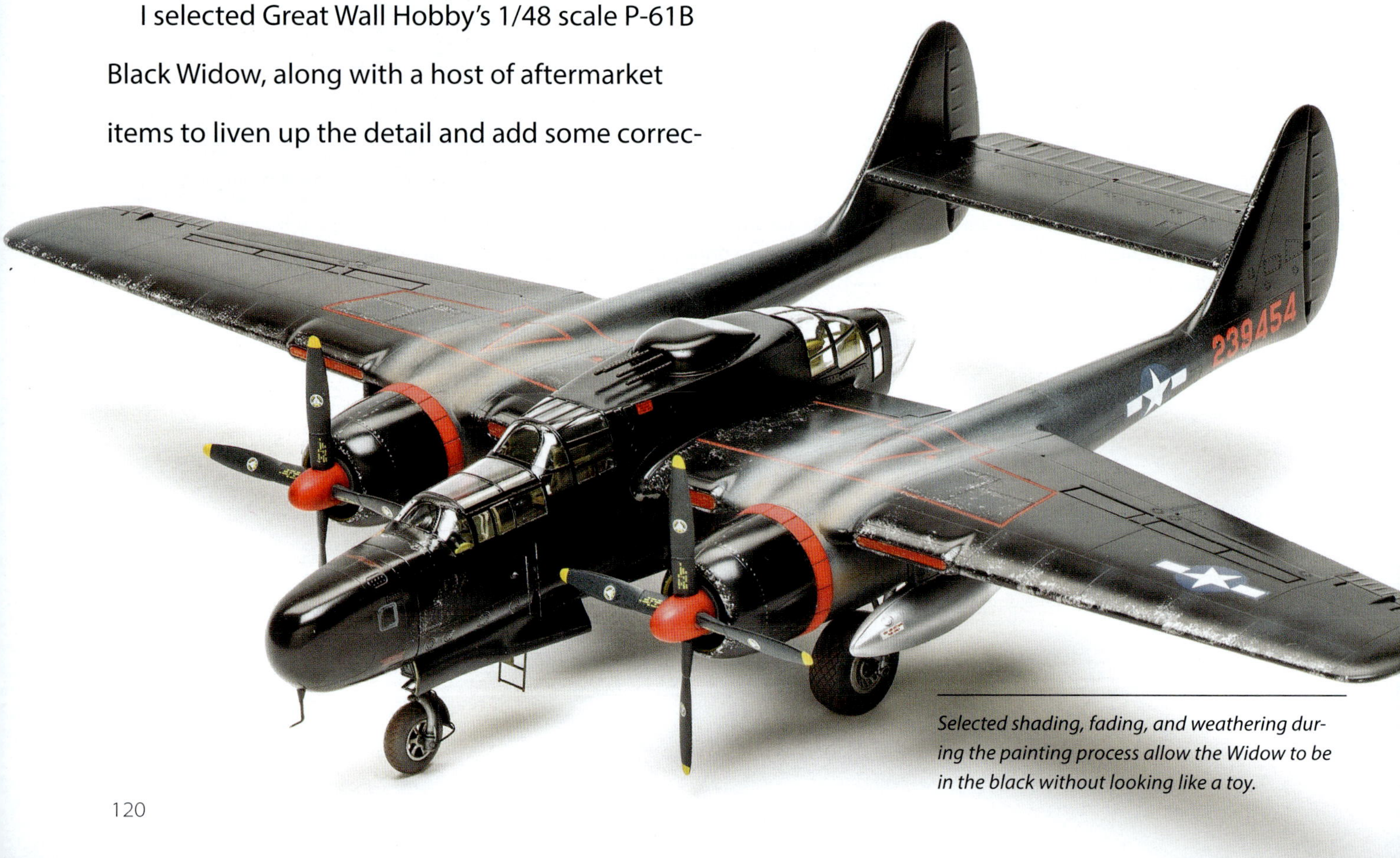

Selected shading, fading, and weathering during the painting process allow the Widow to be in the black without looking like a toy.

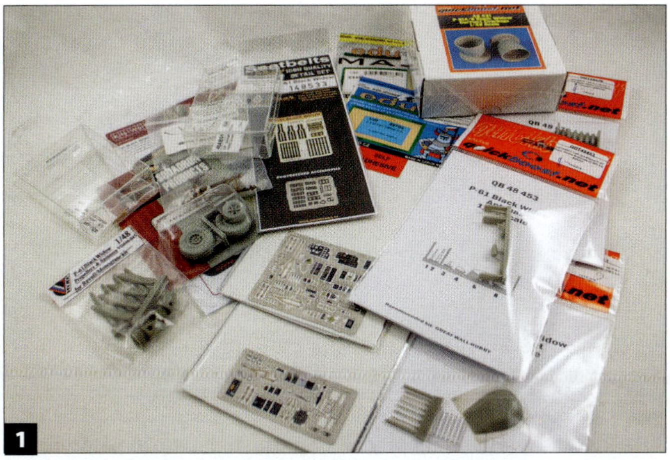

1 Great Wall Hobby's P-61B Black Widow has a lot of nice detail. Of course, there's always more that can be added thanks to Eduard PE and HGW fabric seat belts. In addition to the interior, I added main wheels from Squadron, landing gear from G-Factor, and new spinners and props from Vector Resin. I selected *Cooper's Snooper* from KitsWorld Decals as my Widow of choice which meant I needed a "fixed" top turret available from QuickBoost.

2 Between Great Wall and Eduard, plenty of detail is ready for the interior of the three-place Widow. My PE bending skills failed me when it came to the complicated radio rack in the rear compartment. I substituted a scratch-built rack made from white styrene strip. I left out some of the brass sidewall detail as it seemed too flat.

3 Pre-painted PE installed along with careful painting of the interior leaves the rear of the Widow ready for viewing.

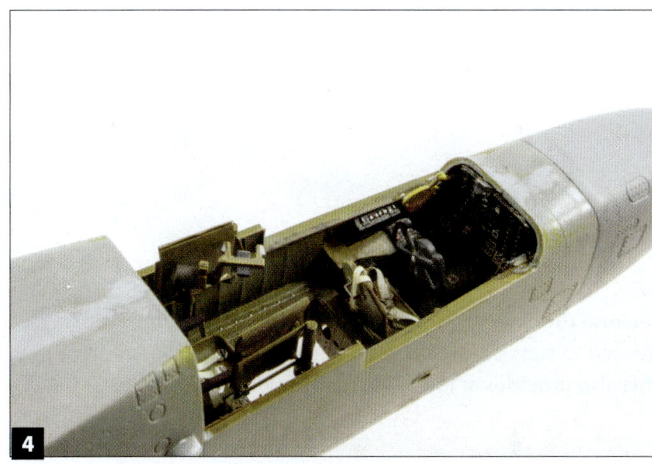

4 The front office has plenty of furniture on display as well—good thing, with all the glass on the Widow.

5 Even seemingly simple paint schemes require all the assemblies to be ready for painting. Since my Widow needed a replacement resin "fixed" top turret, I decided to make a positive mount for it. That requires drilling a hole in the center, which is easily accomplished by using a circle template to find the middle of the turret ring.

6 A couple of quick lines with a Sharpie to mark the center, and I drilled a hole for a plastic rod.

7 And along came a spider. After what seemed like forever preparing the interior, the Widow is ready to start painting. I used Eduard masks—these sets are not only great time-savers, they allow more accurate frames. All other openings were masked with painters tape.

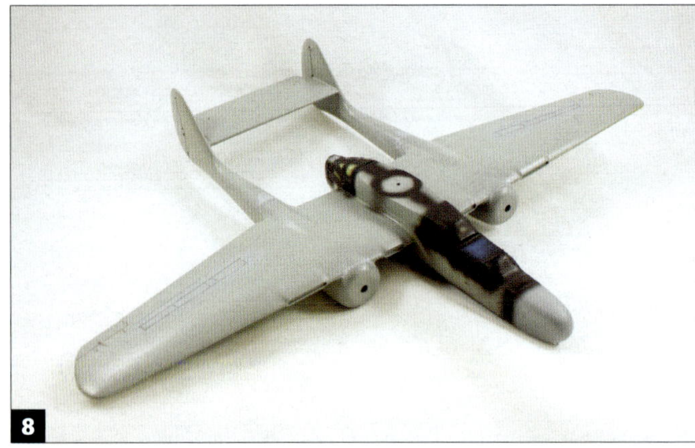

8 It seems like it should be obvious but the first step in painting the Black Widow is to paint … well, black. This allows the inside of the canopy glass to show up as black. While I had the airbrush up and running, I sprayed a couple of the seams to check for hairline cracks.

9 Because there were going to be a number of painting steps, I wanted to start out with a clean canvas. I used Alclad gray primer. This also provides a "faded" undercoat for the next step.

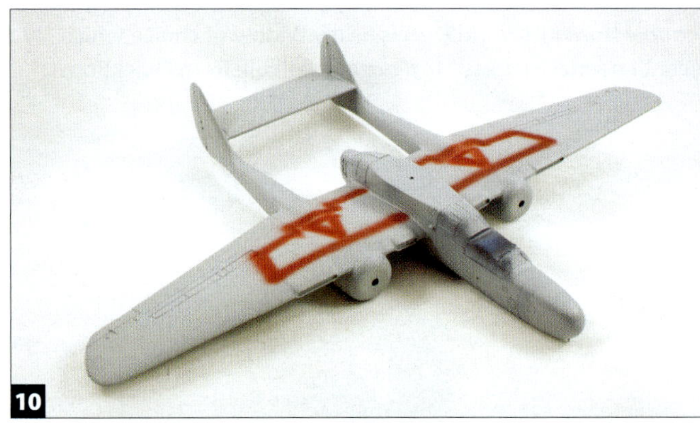

10 How can you tell a Black Widow? By the red markings on its back of course. The kit decals seemed a bit thick and have a reputation for silvering. I couldn't find any stencil replacement decals anywhere. It's easy enough to apply your own wing walk markings though. Simply spray a rough red line in the location of the walkways …

11 ... then use thin pinstriping tape to mask the markings. I always go back over the edges of tape with the back of my fingernail to prevent paint creeping under the tape. Once the markings are laid out as seen on the right wing, I applied more primer to again "even" the canvas (as seen on the left wing).

12 The idea with the painting process is to apply varying shades and areas of paint to break up a monochromatic black finish. I planned to adjust the colors used based on the areas of the model that typically showed the most wear in pictures. For added variety in the final black finish, I pre-shaded a few panels with Tamiya hull red (XF-9). Other than being closer to the engines, there really wasn't any reason for the panels I chose.

13 Next, I used Tamiya dark gray (XF-24) applied in a patchy coat to the wings, booms, and tail surfaces. Pictures show the wings and tail surfaces seem to have the most weathered look compared to the center fuselage and nose areas.

14 Next, metal control surfaces, wingtips, and aft fuselage were painted in Tamiya NATO black (XF-69) as the next least weathered areas.

15 As an added difference in shade, I painted the fabric-covered ailerons and rudders in Tamiya rubber black (XF-85), a slightly blue black. To save time, and wear and tear on the paint, I masked using Post-It notes held next to the edges between colors.

16 The next step requires masking any colors that you wish to remain generally intact—in this case, the control surfaces. Don't forget the associated hinges while you're masking.

17 Finally, some nice gloss black for my Widow. A solid coat of Tamiya black (X-1) is applied to the nose as most pictures I've seen show little wear on the forward fuselage.

18 Next, a thin coat of Tamiya black is gently added over the top of the other colors.

19

While hard to see in the photo, the Widow is not really all black … sort of. The flying surfaces show many shades of black, with a few less on the fuselage. The control surfaces are too stark at this point, but they will be blended in later.

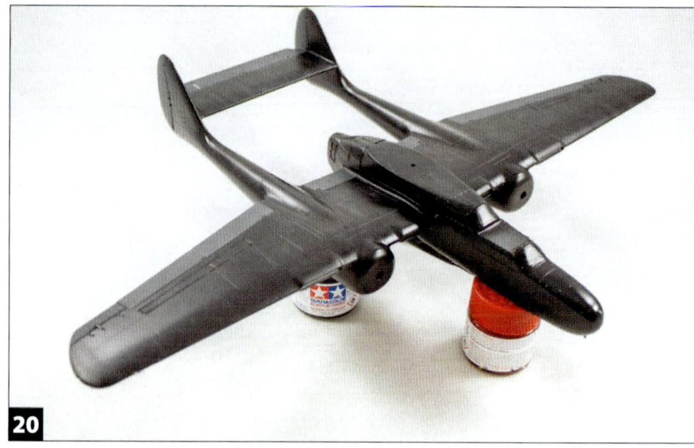

20

Time now to fade and outline panels using Tamiya dark gray (XF-24), focusing mainly on the wings, stabilizers and fuselage underside—areas of highest wear.

21

I added a mix of Tamiya black and brown (X-9) staining near the engines, mainly where the exhaust stains will be placed. Random hazing of a handful of panel lines was done at the same time. You can also see more of the panel lining and random streaking of the previous dark gray.

22

Finally, I applied a thin, streaked coat of Tamiya black over the control surfaces and any areas that seemed to have more contrast than desired, leaving me with a black Widow. But a closer look reveals all is not as black as it seems—there is variation in the colors across the whole model, just what I wanted.

23

Removing the wing walk masking reveals perfect red lines. Once the masks were removed and decals added, I started weathering. Here, differences in sheen are applied using mixtures of Testors Acryl clear flat and clear gloss.

24

When weathering, remember to include the decals along with the rest of the paint. Here you can see streaking and sheen differences across the wing walk areas.

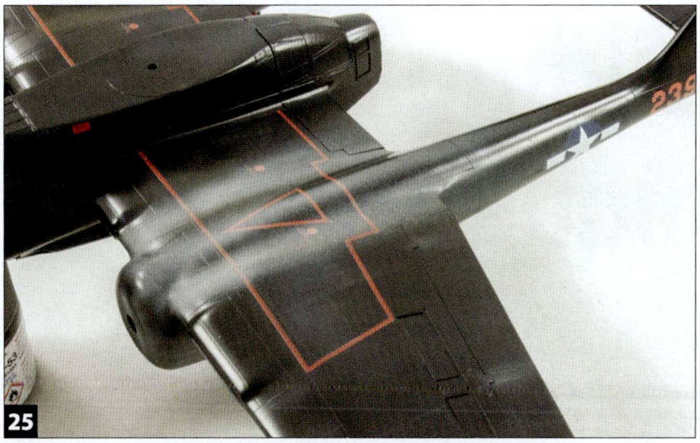

25 The Black Widow had distinct exhaust staining patterns. The first step is applying the lighter, outer portions of the stains using Tamiya deck tan (XF-55). Again, it's important to perform this step with all decals and other markings in place.

26 Finally, the centers of the exhaust stains are … ummm … stained with Tamiya dark earth (XF-52) for a burnt brown.

27 Now that I've spent all this time adding paint to our Widow, it's time to simulate taking paint off. Based on photos, chipping occurred frequently along the leading edges of the wings and stabilizer, less often in other areas. Engine service areas also seemed to take a beating. To add chipping in these areas I used a cut piece of sponge dipped in Model Master steel to dab random spots. The sponge is easier to use than a brush.

28 Not to be forgotten, the engine cowls were treated to all the same painting steps as the airframe. In addition, many photos of Widows show the cowl fasteners often had the paint worn off. This is easily simulated using a silver pencil, adding yet another area of "not black" to our Black Widow.

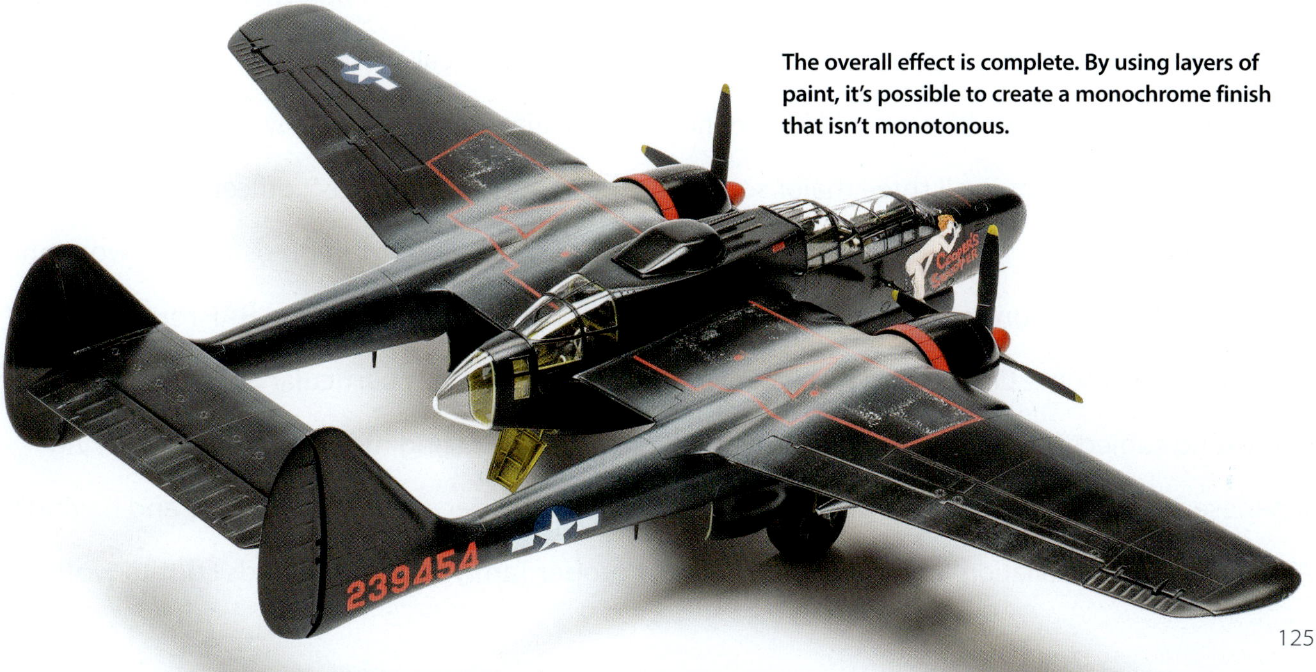

The overall effect is complete. By using layers of paint, it's possible to create a monochrome finish that isn't monotonous.

16 Building a unique Zero

Tamiya's new 1/72 kit meant it was time to tackle the subject again

ANDERS ISAKSSON

Ever since building the old Matchbox 1/72nd scale Mitsubishi "Zero" many years ago, I have had a soft spot for this small and nimble aircraft. There have been many kits of the Zero produced over the years, but in the latest decade we have been blessed with a very fine series of different Zeros in all three major scales from Tamiya.

I was keen to build one of these kits in 1/72 scale, but it took some time until I found an interesting subject for my model. A number of Japanese aircraft ended up at the disposal of ATAIU-SEA, an abbreviation for the Allied Technical Air Intelligence Unit of South East Asia, and was run under the joint command of British and U.S. forces. The purpose of this unit was to evaluate Japanese aircraft and their capabilities from both a technical and tactical view. Interestingly, these aircraft were flown by Japanese pilots under the supervision of British officers.

I spotted a photograph of a rather worn and weathered A6M2b wearing RAF roundels and I realized that this particular aircraft would make for an interesting modeling subject. Having found the decals for the machine and having the proper Tamiya kit in my hands I got to work.

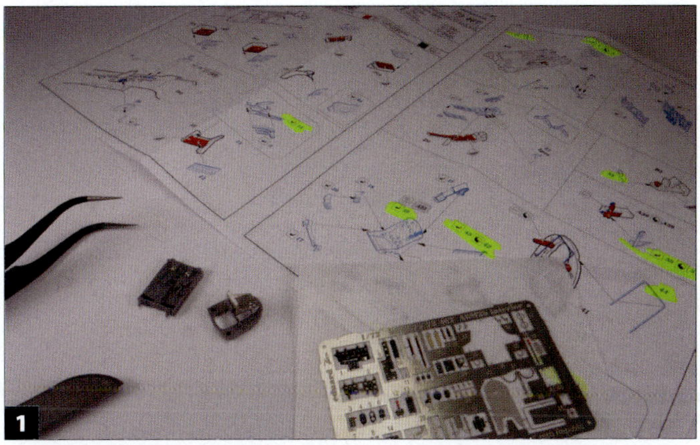

1 Careful planning to sort out when to add specific photo-etched metal (PE) details helps construction. All parts that can be added before any painting is done were marked on the instruction sheet in yellow highlighter to ensure no details are forgotten. For me, the challenge with using PE details in the cockpit is not so much the actual construction, but more to figure out the logical assembly process. This gets more complicated when the PE parts comprise a mix of pre-painted and unpainted parts. Figuring out the order of assembly is crucial to ensure everything gets painted the correct color while minimizing the need to change colors during painting.

2 The cockpit and fuselage sub-assemblies are ready for painting. At this stage only the unpainted PE parts have been installed.

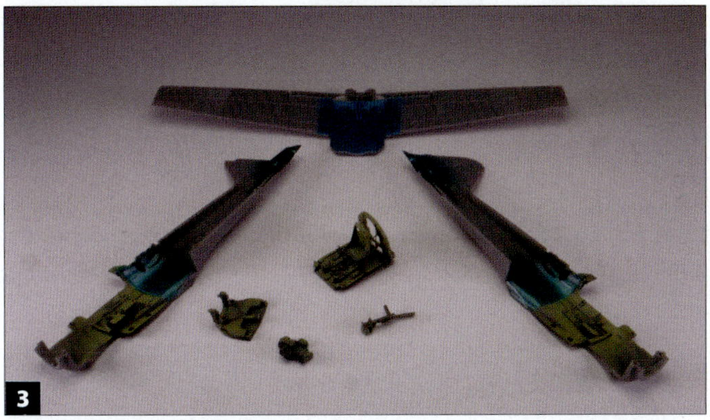

3 Once the unpainted PE parts had been attached, it was time to apply the overall cockpit color, which in this case was the Mr. Color Nakajima cockpit green (127). Also, a few specific areas of the interior were painted in the peculiar Japanese aircraft primer called Aotake. This is a blue-green metallic shade and for this I used Mr. Color metallic blue green (57).

4 The engine received a PE ignition harness, which makes a notable improvement. I accidently used the wrong crankcase, but I exchanged it later for the correct version. Depending on whether the part needs to have a strong bond or not, I use either regular superglue or Gator's Grip water-based glue. Removing burrs from the PE parts results in small areas of bare metal being exposed on the pre-painted areas, but this is best touched up with a fine paintbrush once the parts have been installed.

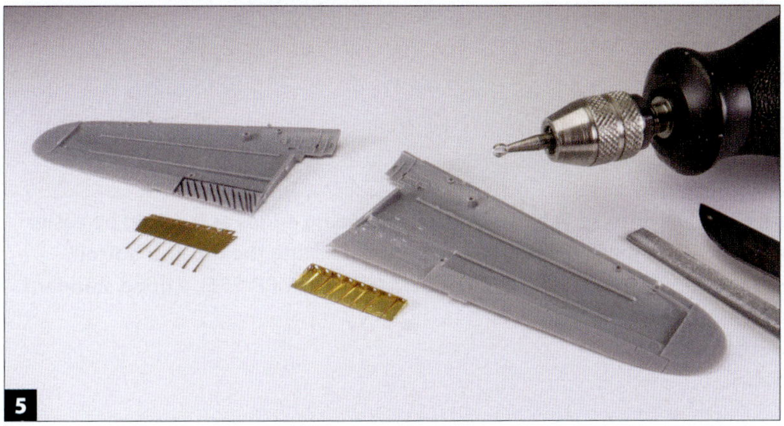

5 Once the cockpit was installed, it was a breeze to get the rest of the airframe together. I deviated from the instructions with the main wings where I modified the flap area and removed large portions of the kit plastic to make room for the PE flap interiors. Using my trusty rotary tool, I thinned the insides of the upper wings until the PE interior inserts fit as intended. The flap interiors were relatively easy to assemble and install using superglue. The flaps were likewise rather easy to construct using minimal amounts of superglue to hold them together. Their interiors and exteriors were painted and the flaps kept separate until final assembly.

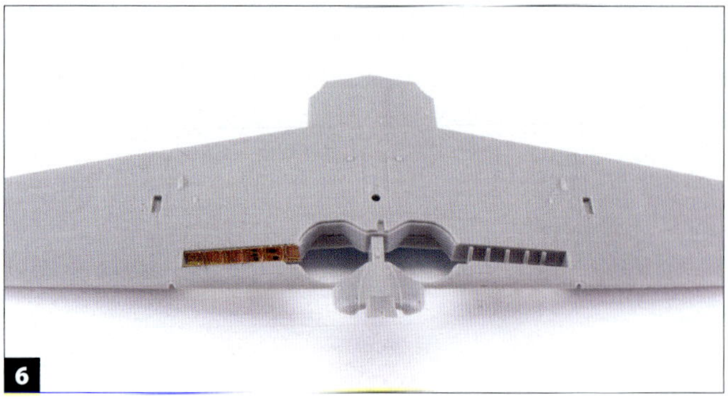

6

The wheel wells were also enhanced with PE details. The bulkheads were the major improvement. Note the difference between the finished wheel well on the left and the original well on the right.

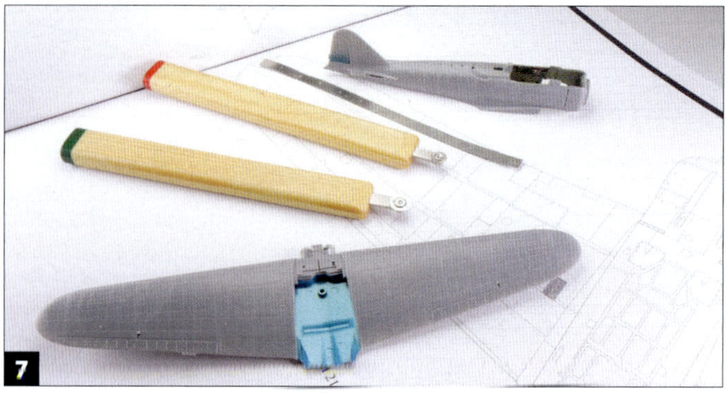

7

To further detail to the exterior, I added rivets to the entire airframe using my Rosie The Riveter tools. My guide was the excellent drawings in *Mitsubishi A6M Zero* by Dariusz Karnas (ISBN 978-8363678364). I did not follow the drawings down to every single line of rivets, preferring the overall effect of rivets. Also, I made sure to add most of the rivets to the wings and fuselage while they were separate—it was easier.

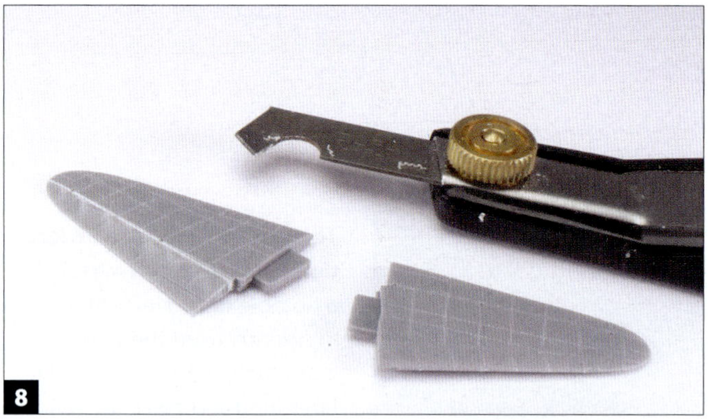

8

By carefully scoring the underside of the elevator hinge lines, I was able to position the elevators with a realistic droop, adding personality to the model.

9

After wiping down the entire model with denaturated alcohol, I applied an overall coat of Mr. Paint black primer (MRP-085).

10

When the primer was dry, I applied a coat of Mr. Paint white aluminium (MRP-009) to form the base for exposed metal surfaces. Using small pieces of sponge, I dabbed small amounts of Ammo by Mig Jimenez masking fluid around the airframe, using the photograph as an inspiration for the weathering pattern. All control surfaces were masked off as these were covered with fabric and should not show exposed metal.

11

I applied yellow to the wing leading edges as it would be easier to mask these areas off before painting the top and bottom colors. I used Mr. Paint RLM 04 gelb (MRP-051), a good standard yellow.

Next up was the typical IJN green camouflage color, and here I went with Mr. Color's IJN green (15), lightened with a bit of Nakajima cockpit green for a slightly weathered appearance. Once the green dried, I carefully polished the model with a piece of dry tissue paper, which helped remove the masking fluid, exposing the aluminium underneath.

The control surfaces were masked and painted a slightly lighter shade of green to replicate faded fabric. To add more contrast to the overall green color and stay with the photograph I used as inspiration, the rudder and elevators were painted the original IJN green, as these components seemed to be a distinctly darker shade and perhaps had been taken from a newer aircraft or were fresh spares.

While I had the original IJN green out, I re-created the traces of the overpainted national insignia on the fuselage. I roughly cut out a pair of circles from pieces of Tamiya tape and carefully airbrushing inside each mask on either side of the fuselage.

Moving to the undersides I masked off the upper surfaces, making sure to make the demarcation line between the gray and the green along the rear fuselage curve up against the elevator, a tell-tale sign of this being an airframe built by Nakajima as opposed to Mitsubishi. The wheel wells were masked off with pieces of foam before applying the gray. I left the flap wells unmasked as I find it is easier to mask around them and apply the interior color afterward. For the entire undersurface I applied a coat of Mr. Color IJN gray (35). Once the gray dried, I masked and painted the flap wells, again using the standard Aotake primer.

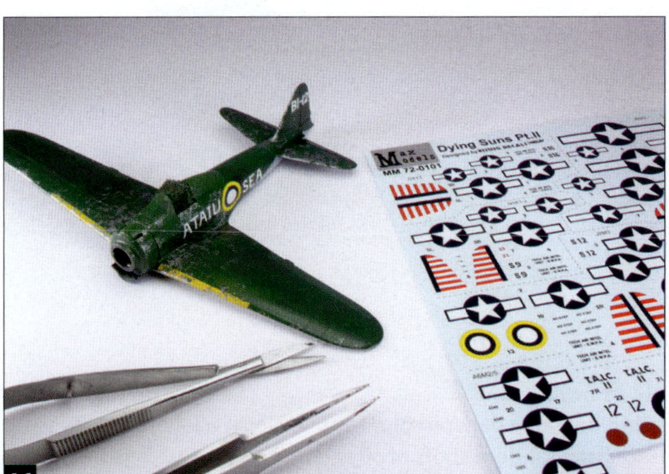

The decals from Max Models worked very well. All of the original national insignia on this particular aircraft had been painted over and replaced with British-style roundels. Only small amounts of Microscale Micro Sol applied in selected areas was needed to make the decals conform to irregular surfaces. I refrained from applying stenciling as I figured the real thing was hastily painted with little effort spent on the finer details.

17

The only other marking I applied was the rather prominent red lines on the upper wings. These were created by masking them with tape and painting them on using small amounts of Vallejo flat red (70.957) and a small piece of sponge along with some careful painting with a fine brush. Once everything had dried, I applied an overall coat of Mr. Paint clear flat (MRP-127) to seal the markings and prepare all surfaces for the next phase.

18

I continued by applying an overall wash of diluted oil color using starship filth from Ammo. This is a dirty dark gray color that adds depth as well as to represent collected dirt and grime. Working in sections, I applied the oil color and let it dry, then wiped the area with a piece of lint-free paper, leaving the oil color in panel lines and other recesses. Here the difference between the treated and untreated (farthest from the camera) lower wing can clearly be seen.

19

Although not too readily apparent in this photo, light brown shade was applied with a sponge to simulate worn propeller blades and spinner, creating variation and further interest to this area.

20

The next step was add some further variation to the green upper areas by applying tiny paint chips and scuff marks using a few different artist pencils. Finally, a few selected areas of bare metal were enhanced with a silver artist pencil.

21

A heavier application of dark oil color was streaked along the lower center wing section, simulating traces of leaking engine oil along with general dirt and grime.

22

Using the same dark oil color I applied local washes to replicate traces and random spots of oil and dirt on the fuselage behind the engine.

23

After installing the landing gear, I added some dry mud and dust to the wheels by applying a few shades of Mig Productions pigments with an old brush.

25

At this stage I carefully moved the model back and forth across a sheet of sandpaper to create flat spots on each wheel to simulate the weight of the aircraft.

24

The instruction for the decals shows this Zero being armed with one Type 99 20mm cannon in each wing, so I installed a set of turned-brass barrels from Master. I submerged the barrels in AK Interactive's Metal Burnishing Liquid for a few minutes, giving them a dark finish without the need for painting. The barrels were installed with tiny drops of superglue, then rubbed with graphite powder from a lead pencil to mimic the look of gun metal.

26

To wrap it up, I added smaller stuff like the pitot tube, tail wheel, and antenna mast that goes through the canopy behind the cockpit. For the antenna wire I used a length of fine flexible rigging thread from Uschi Van Der Rosten, attached at both ends with tiny drops of superglue. Finally, the flaps and propeller were installed and there it was—my model of a captured Mitsubishi Zero in rather unusual markings was complete.

17 'Hello again!'

Monogram's 1/48 scale Devastator was the basis of a scene set 6,500 feet below the ocean's surface

DARREN ROBERTS

A video released when the Lexington was found showed a full 360 degree view of a lone Douglas TBD-1 Devastator resting upright on the ocean floor. There was very little marine growth, and not much going on "action-wise". Yet, it was the starkness of the scene that made it intriguing. This is what I set out to replicate.

As a history buff as well as a modeler, I was fascinated by the discovery of the USS *Lexington* by the research ship R/V *Petrel*. Sunk on May 8, 1942, during the Battle of the Coral Sea, "Lady Lex" sat undisturbed on the ocean floor for nearly 76 years, seemingly lost to history. But on March 4, 2018, the crew of R/V *Petrel* recorded astounding images that brought history back to life in full color. At more than 6,500 feet down, the *Lexington* showed very little deterioration because of the cold temperature and lack of oxygen in the water. As more and more images and video were released, the sense of historical awe and wonder grew exponentially. Never would I have imagined I would see such a historical treasure in such good condition. It was the ghostly images of the lost aircraft, however, that spurred my imagination the most and inspired me to try my hand at a diorama.

I procured a 1/48 scale Monogram Devastator from a friend. I had forgotten what a beautiful kit this was, even though it was more than 40 years old. A funny thought occurred to me as I was looking over its contents. There was less time between when the *Lexington* was sunk and when Monogram released the kit (32 years) then there was between when the kit was released and today. Isn't it strange how time moves so fast? But I digress! It was time to make a diorama.

Though in relatively good condition on the ocean floor, there was still damage to the Devastator. The most striking was the rudder, which had broken away from its post and was partially de-skinned. I started with this, as I thought it would be the most difficult to replicate.

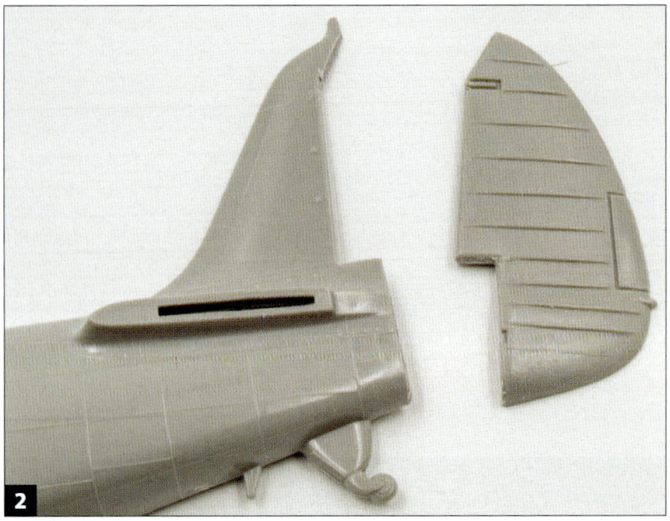

I separated the rudder from the tail using a scribing tool to thin the plastic.

From there, I removed the rudder hinge ...

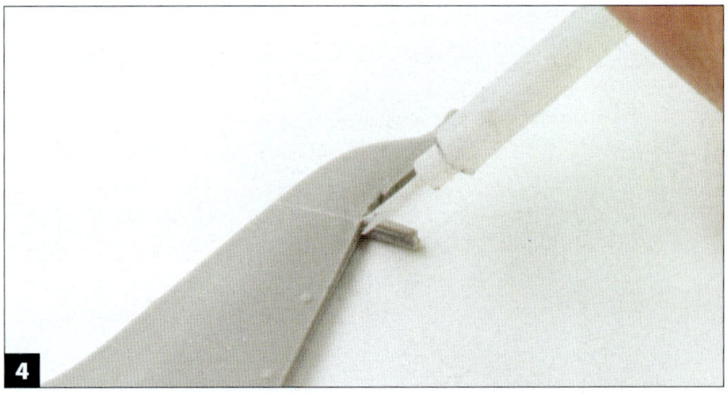

... and glued it back on the tail.

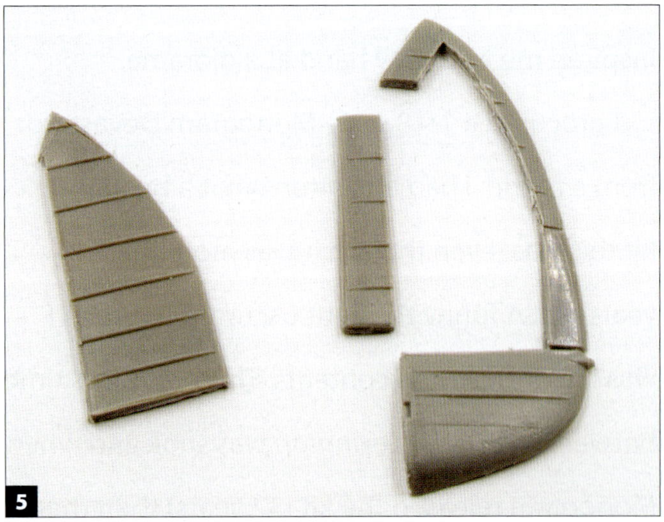

I cut the plastic from the center of the rudder, leaving just the outline.

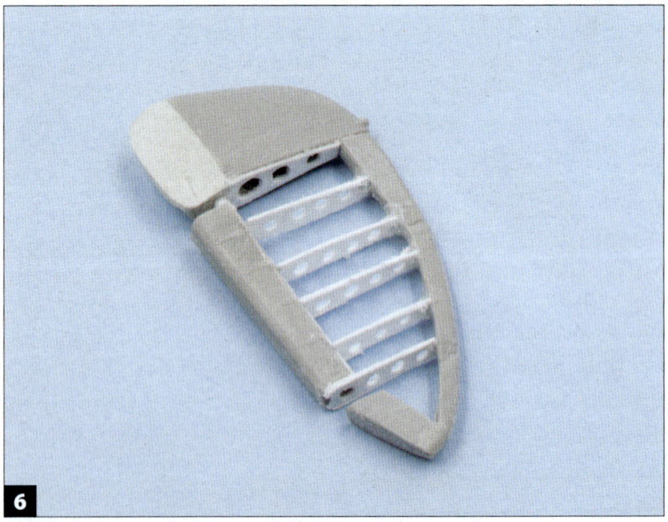

Using strips of thin styrene plastic, I fabricated ribs and glued everything together. The lightening holes were drilled before I assembled the parts. Once the ribs were glued in place, I sanded them to the correct shape. I finished off the form of the rudder by using two-part epoxy putty to fill in the missing base of the rudder.

7

To replicate the skin coming off of the rudder, I used aluminum foil from my kitchen. It's thin yet strong, and would hold its shape. I glued it to the outer edge of the rudder frame using superglue and trimmed off the excess.

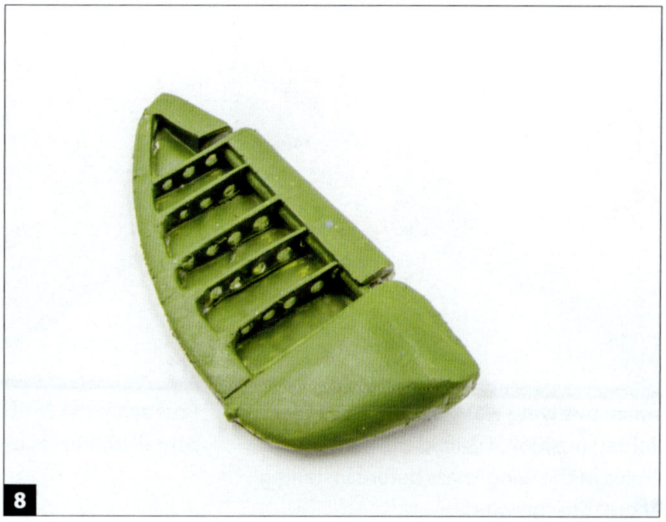

8

I sprayed everything with interior green before I covered up the other side. The foil for the opposite side was painted green before I glued it on.

9

I primed the rudder using Alclad gray primer to ensure decal adhesion. When it was dry, I applied, the rudder stripe decals.

10

When the decals were dry, I used tweezers to pull away the foil from the ribs. The finished assembly was set aside to be attached later in the build.

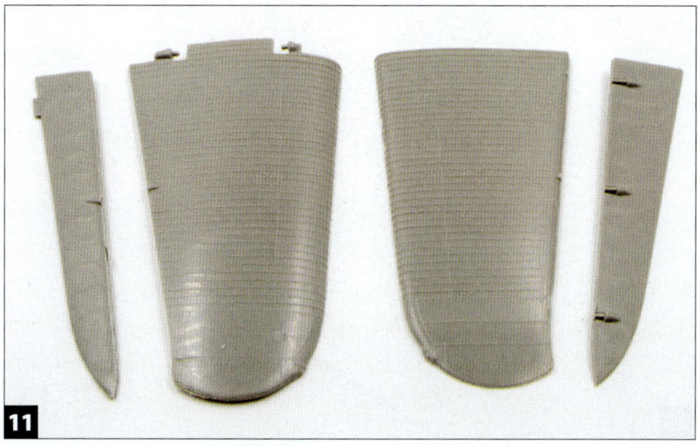

11

The next step was repositioning the control surfaces on one wing and on both horizontal stabilizers. I cut away the aileron from the right wing using the same technique as the rudder.

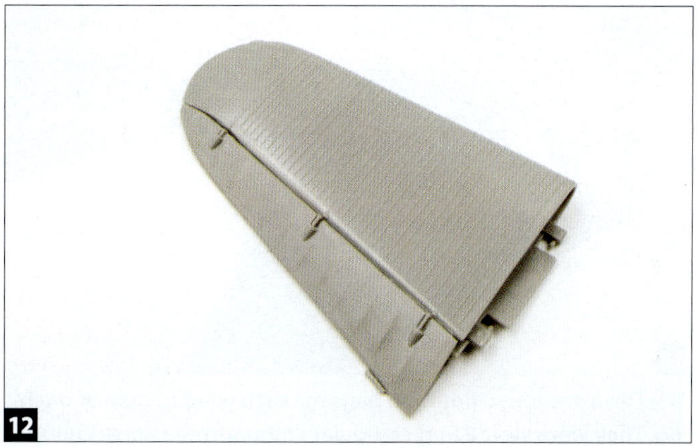

12

The wing halves were then glued together and the aileron was put in place.

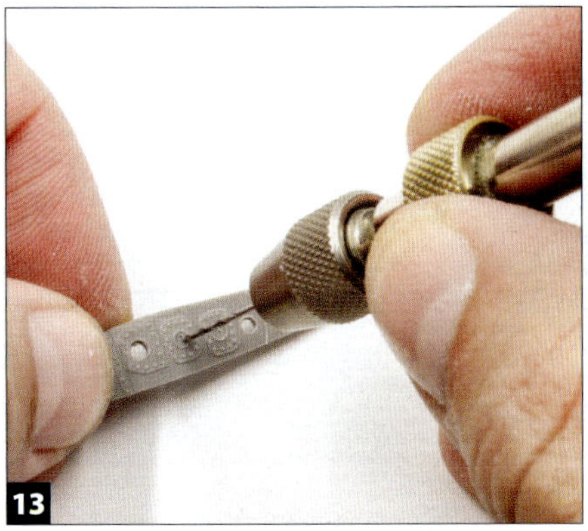

13

Since this wing was going to be in the folded position, I drilled the lightening holes in the wing spars before installing them into the wing.

14

I cut away the control surfaces of the horizontal stabilizers using a hobby knife.

15

The same tool was used to remove the hinges.

16

A flat metal file finished the job.

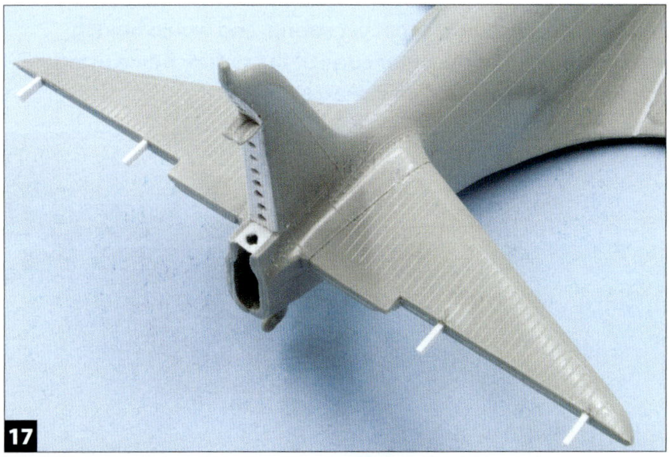

17

I used strips of styrene plastic to replace the hinges. They were glued back on to the horizontal stabilizers in the correct position. I also added sheet styrene spars to fill the opening left after removing the rudder.

18

The Devastator had flotation bags for each wing in case of ditching. They were stored in rectangular compartments near the fuselage toward the leading edge of each wing. In the images of the TBD I was modeling, the compartment on the left wing was open and the bag was missing. To replicate the open space, I used masking tape to mark off the area that needed to be removed. The panel lines of the kit show where to cut. I used a scribing tool to thin the plastic.

19

I glued sheet styrene to the underside of the wing to box in the opening. Did the bag detach as the aircraft sank? Did one of the crew of the *Lexington* open it up and try to use it as a life raft? It's just another intriguing question that stirs the imagination about this long-lost airplane. The assembly was set aside to be put on later.

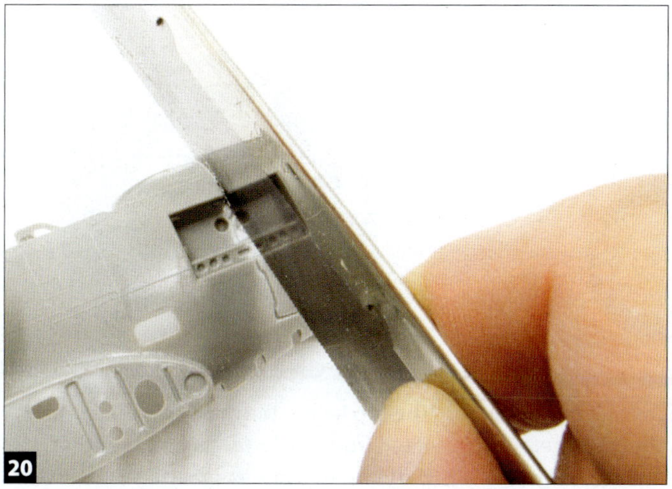

20

Now I moved on to the fuselage. There were two areas that needed to be worked on: the engine and the empennage. I tackled the engine first. The top access panel was missing on the full-sized plane, so I cut that away using a razor saw.

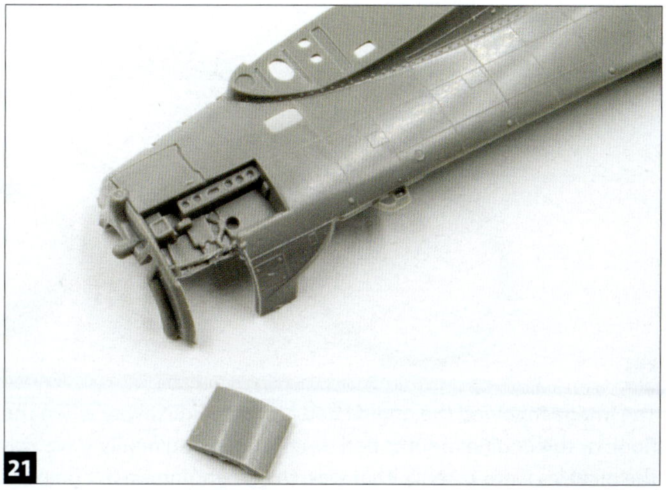

21

This area contained a fuel tank and associated fuel and hydraulic lines.

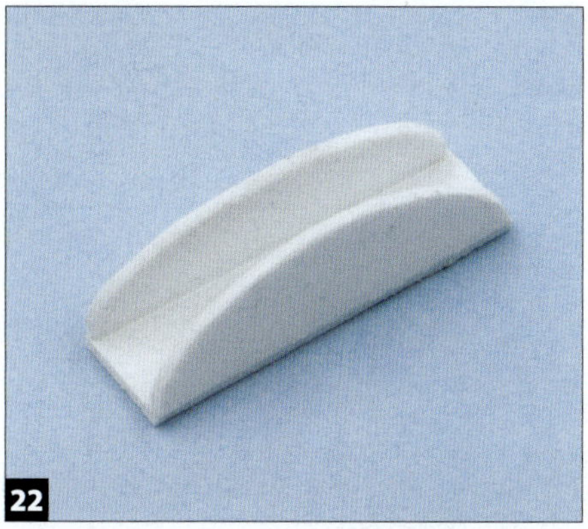

22

The fuel tank was fashioned from sheet styrene cut to shape.

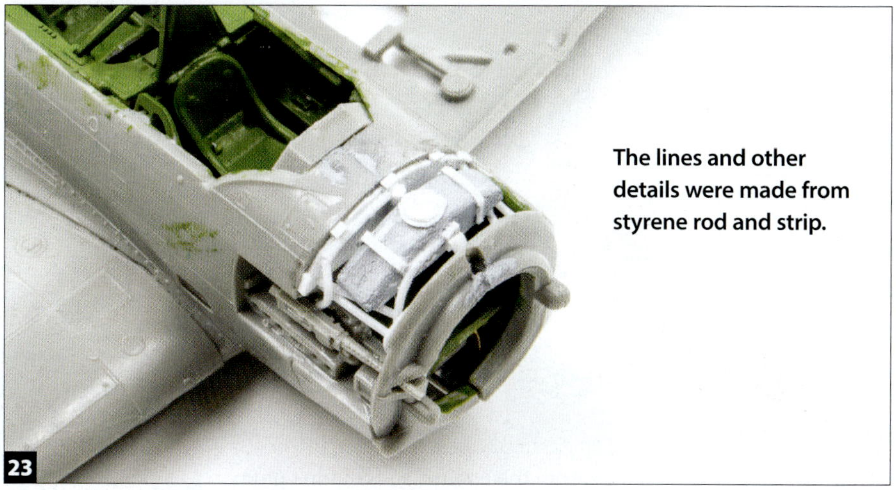

23

The lines and other details were made from styrene rod and strip.

24

I used a pin vise to open up the lightening holes for the machine gun mount ...

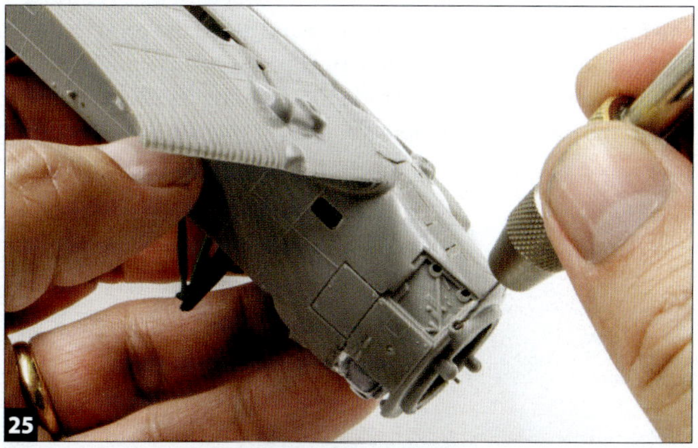

25

... and the exhaust ports.

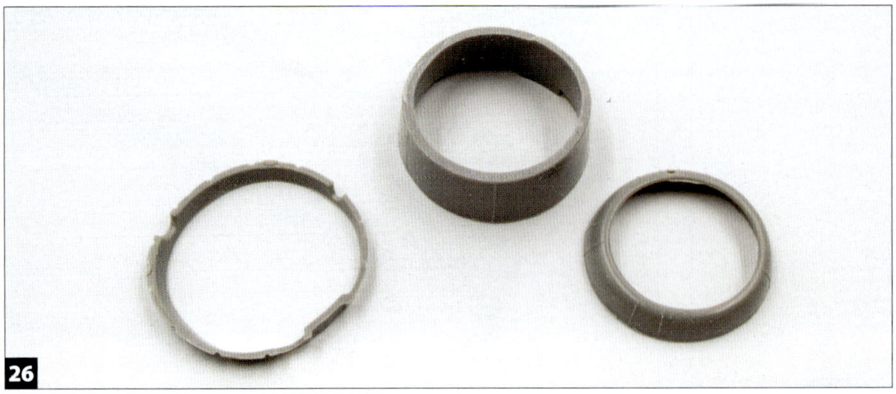

26

The images showed the engine had either broken away when the aircraft hit the ocean floor, or the engine mounts had deteriorated and finally gave way in the years before the pictures were taken. Either way, the kit engine wasn't going to fit the bill. Vector Resin makes a beautiful Pratt & Whitney R1830 engine with complete rear detail, which would be visible given that the cowling was pulled away from the front fuselage. To prepare for the resin engine, I used a razor saw to cut the kit cowl into three parts, separating the cowl flaps, the main cowl, and the front ring.

27

The top half of the main cowl had separated from the lower half, so I cut that part into two pieces. I reattached the cowl flaps to the lower half of the main cowl, painted it, and secured the completed resin engine with superglue.

28

The propeller had broken free from the engine, so I couldn't glue the prop directly to the engine shaft. To deal with this, I painted the front cowl ring and the propeller and glued the prop to the ring to keep it in place. I painted the upper cowl half and reattached the front ring to it, setting the assembly aside until the rest of the model was completed.

29

I finished construction on the rest of the model, and then went to work on the broken empennage. Using a razor saw, I cut the rear area at an angle.

30

I stopped cutting as I neared the bottom of the fuselage and removed the empennage with brute force. Doing it this way created an uneven edge.

31

This was followed by thinning the plastic on the inside of the fuselage pieces and creating jagged edges with a hobby knife.

32 The empennage was then glued back on, but at a bit of an upward angle to represent the broken structure. When the joint was completely dry, I used strips of Bare-Metal Foil to cover the gap. This allowed me to represent the wrinkled metal skin as well as being able to create holes and tears.

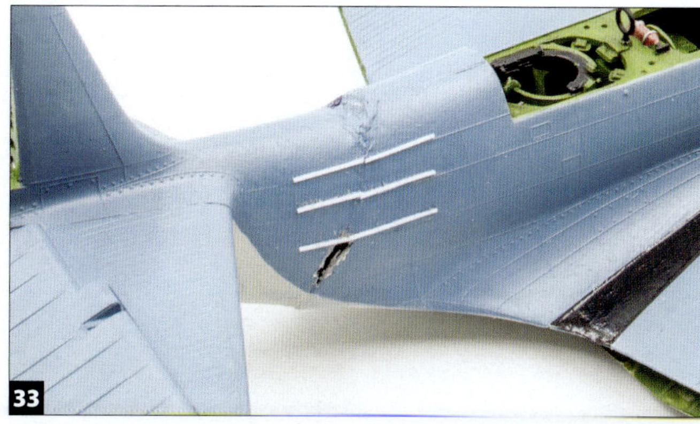

33 With the empennage attached, I finished the rest of the model, using Model Master Acryl paints for the blue-gray on the top and light gull gray for the underside. I was about to start decaling when I noticed that on one side of the fuselage where it had broken, you could see the shape of the stringers. I glued .25mm rod to the kit's surface, then sanded them down to blend them in.

34 After repainting that area, I added the decals,

35 and attached the rudder,

36 and engine.

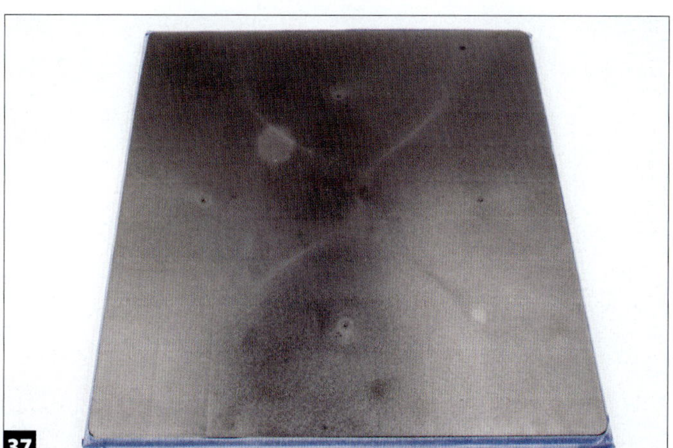

37 Now it was time to make the base the model would sit on. I used a display case from Trumpeter made for 1/48 scale models. It has a black plastic base and clear plastic cover. The images of the Devastator show the sea floor to be smooth and sandy. To replicate that, I masked the edges of the base.

38 Then I sprayed the surface with Rust-Oleum Multi-Colored Textured Surface spray paint. To get depth, I emptied the can and let the paint build up. When the paint was still a bit squishy, I pushed the Devastator into it to avoid it looking like it was sitting on a hard surface. I removed the model, and after the paint was dry, I glued it back into the depressions with superglue.

39

Next came the small details. The engine materials had corroded, making what looked like a pile of rubble in front of the airplane. Crossing into another modeling genre solved the problem. Model railroad ballast, which is used for the bed of the track, was a perfect fit. I used both medium and small sizes. I poured white glue where I wanted to the ballast to adhere,

40

then dumped fingers full of ballast on top of that. This was done in multiple layers, with the first couple of layers being the larger sized ballast and the last few layers using the smaller sized ballast. The excess was brushed away into the trash can.

41

Another model railroad material helped out when it came to replicating the marine life that had grown on the surface of the TBD. Fall foliage ground foam had the right look, and it was easily applied. I used a small brush to dab white glue on random areas of the model.

42

I then set bits of the foliage on top of that. The material stuck almost instantly to the white glue, but I was still able to move it around and position it where I wanted.

43

For the mushroom-looking things (Hey, I'm a fourth-grade teacher, not a marine biologist!) I drilled small holes at each location.

44

Then I glued a bent piece of styrene rod for the stem.

45 For the tops, I made a small "snake" of two part epoxy putty, cut it into small bits ...

46 ... and attached it to the top of the stem with superglue.

47 I wasn't happy with the color of either the ballast or the foliage, so I carefully painted them. The ballast received a wash of gunship gray.

48 While the foliage was overpainted with light gull gray to tone down the yellow.

49 To give more variation to the ocean floor, I added some of the larger-sized ballast ...

50 ... and painted it engine gray.

51 I also dry-brushed light gull gray mixed with light gray in random areas to create some tonal differences.

I added some "junk" around the model to busy it up just a little bit, and I called it good.

52

53

But something wasn't quite right. This was supposed to be an underwater diorama, but with the clear plastic cover in place, it looked more like a beach scene.

How could I get that dark blue, deep ocean look? I thought of using clear resin, but I wasn't bold enough to try it. My next idea was to simulate it with lighting. At least if that failed, all I had to do was purchase a new clear top. I started by spraying the insides of the clear cover with Alclad clear blue.

Then it was off to my local automotive parts store, where I picked up a roll of window tint. I cut pieces that were approximately the same as each of the sides of the cover and stuck them to the outside of the cover. These were then trimmed to get a nice fit.

This project allowed for a different take on aircraft modeling and diorama construction, and even without the cover, makes for an interesting display.